The Complete Cat

www.rbooks.co.uk

Also by Vicky Halls

Cat Confidential
Cat Detective
Cat Counsellor

For more information on Vicky Halls and her books, see her website at www.vickyhalls.net

THE COMPLETE CAT

VICKY HALLS

BANTAM PRESS

LONDON • TORONTO • SYDNEY • AUCKLAND • JOHANNESBURG

TRANSWORLD PUBLISHERS
61–63 Uxbridge Road, London W5 5SA
A Random House Group Company
www.rbooks.co.uk

First published in Great Britain
in 2008 by Bantam Press
an imprint of Transworld Publishers

Copyright © Vicky Halls Ltd 2008

Vicky Halls has asserted her right under the Copyright, Designs
and Patents Act 1988 to be identified as the author of this work.

This book is a work of non-fiction based on the life, experiences and recollections of the author. In some limited cases names of people, places, dates, sequences or the detail of events have been changed solely to protect the privacy of others. The author has stated to the publishers that, except in such minor respects not affecting the substantial accuracy of the work, the contents of this book are true.

While all reasonable efforts have been made to ensure that the behavioural advice and information provided in this book is accurate, it is not a substitute for expert veterinary advice. Animals should always be taken to their vet in the first instance.

A CIP catalogue record for this book
is available from the British Library.

ISBN 9780593061121

This book is sold subject to the condition that it shall not, by way of trade or otherwise, be lent, resold, hired out, or otherwise circulated without the publisher's prior consent in any form of binding or cover other than that in which it is published and without a similar condition, including this condition, being imposed on the subsequent purchaser.

Every effort has been made to obtain the necessary permissions with reference to copyright material, both illustrative and quoted. We apologize for any omissions in this respect and will be pleased to make the appropriate acknowledgements in any future edition.

Addresses for Random House Group Ltd companies outside the UK
can be found at: www.randomhouse.co.uk
The Random House Group Ltd Reg. No. 954009

The Random House Group Limited supports The Forest Stewardship Council (FSC), the leading international forest-certification organization. All our titles that are printed on Greenpeace-approved FSC-certified paper carry the FSC logo.
Our paper procurement policy can be found at
www.rbooks.co.uk/environment

Typeset in 12/15.75pt Cochin by
Falcon Oast Graphic Art Ltd.
Printed and bound in Great Britain by
CPI Mackays, Chatham, ME5 8TD

2 4 6 8 10 9 7 5 3 1

To Mangus, my muse

LONDON BOROUGH OF HACKNEY	
913 000 00022161	
HJ	20-Nov-2009
636.8	£14.99

CONTENTS

ACKNOWLEDGEMENTS

This book would not have been such a pleasure to write without Clare Hemington. Her comprehensive research, support and good humour have been invaluable. I would also like to thank Sharon Cole for teaching me the art of brevity when discussing veterinary matters; there really is a booklet inside you, Sharon, just waiting to be written! Thanks also to Professor Danielle Gunn-Moore and Dr Sarah Caney for checking the technical information in Appendix 2; I will always appreciate your friendship and wisdom.

As ever, Mary Pachnos, my agent, and Francesca Liversidge, my editor, have been wonderful. Thank you for continuing to laugh in all the right places.

Enormous gratitude goes to all my friends and family, especially Steve, Suzanne, Grace and Annabel Pattle, Sharon Maidment, Nick Murphy, Peter Halls, Mel Reid, Rita Harris, Ruth Yates-Boulton, Pat Shoebridge, Amanda Riley and Janet Valentine, who once again have accepted my disappearance from all social contact while writing *The Complete Cat*. You really are the best in the world.

Last but definitely not least, thank you, Charles, for your love and support; you don't know how much it means to me.

INTRODUCTION

SO MUCH HAS HAPPENED SINCE I WROTE MY FIRST BOOK, *Cat Confidential,* in 2002. I have been fortunate enough to write two more (*Cat Detective* and *Cat Counsellor*), continue with my work as a cat behaviour counsellor and enjoy many other privileges afforded to those with a reputation for being a 'cat expert'. Ironically, I have never described myself as such, because that implies I know it all; on the contrary, I don't think I will ever know enough! Every day those little feline creatures teach me something new, which only goes to show that not all cats respond the same way in any particular situation. You can generalize and say that some will do 'this', while others don't, but the frustration (or the fascination, depending on how you look at it) is they are all individuals and their response to life will depend on their lineage and their experiences.

The most extraordinary and unexpected thing about writing these books on cat behaviour has been the thousands of new friends from all over the world I have discovered. Cat lovers have been extremely generous by writing or emailing me to say how much they have enjoyed the books. Many of them have added a query or two about their own cats that I will endeavour to answer to the best of my ability. Over the past few years these questions have started to include other conundrums that can't really be placed under the umbrella 'behaviour'. The clients from my behaviour practice will also telephone me to discuss practical problems, such as 'Which cattery should I choose?' or 'Is this grooming brush going to be right for Ginger?' I've had numerous heart-wrenching conversations about medical diagnoses and sudden illnesses when owners feel, rightly or wrongly, that I am the person to break down all the complicated terminology and give them the bottom line on their cat's condition. I'm very flattered and, after a brief chat, usually refer them back to their vets, but I'm beginning to think that specializing in behaviour is only half the story. This book, therefore, is something a little different for me. It's still my voice, my words, my thoughts, but with a great deal of research and knowledge acquired from working with experts in the fields of veterinary medicine, general care and behaviour whom I am privileged enough to call my friends.

Just one more thing: for ease of language I refer throughout this book to cats in the masculine. This shows no intended bias – my own cats, Bink and Mangus, are both female.

CHAPTER 1

How Much Is That Pussy in the Window?

ACQUIRING A CAT

MANY OF THE POTENTIAL PITFALLS OF CAT OWNERSHIP ARE experienced by the unprepared. I have spoken over the telephone to so many new owners who have subsequently remarked, 'If only I had known that before I went ahead.'

When, as a newlywed, I decided the time was right to acquire my first cat I spent hours at the library researching breeds, cat care and cats in general to ensure I would a) be a suitable owner and b) choose the cat that was right for me. I'm glad I did, because it prepared me well and every subsequent moment with my cat companions over the years has been a pleasure.

You may be considering an addition to your existing cat family or even embarking on a new cat relationship having enjoyed one or more in the past. You already know the pleasure involved in cat ownership and the decision to go ahead has probably already been made. You may even be a member of a very special group of people I like to refer to as 'cat wannabes'. If you have never owned a cat before or never had a cat as a child I would like to address you specifically for a brief moment.

Hello,

Let me tell you what a cat will bring to your life.

A cat is a naturally undemanding creature that usually takes out of the relationship what you are prepared to put in and nothing more. Cats have moments when they seem intensely pleased to see you and yet they don't sit and pine when you are out. They have a secret life outside your relationship, but when you are there you will be appreciated. They may show love to you when you are busy and spurn you when you want a cuddle; it's not always possible to turn their love on like a tap.

Cats seek out warm, secure places to sleep and so often prefer to share your bed. They see this as a privilege and not a right, so a hammock attached to a radiator or a place near the Aga will do instead.

If you are sad they will fall off the back of the sofa for you or lie with their legs in the air just to make you laugh. They are comedians in fur coats.

They will eat expensive prawns or simple cat food; the trick is to start as you mean to go on. They accept their lot – but the more you give, the more they will want. Be brave enough to have rules and you should still manage to maintain control of the household.

You will struggle with giving them pills, hate car journeys with them, be revolted by their hairballs and have every comfy chair in the house covered with a cat blanket. What you will get in return from a cat, if you are lucky, is twenty years of shared experiences, house moves, growing children, heartaches and happiness. You will take a million photographs of him over the years and, eventually, you will develop your own personal language to understand what he wants and when he wants it. When he is gone you will mourn and marvel at how much you could miss such a small furry creature.

In conclusion, I am recommending wholeheartedly sharing your home with a cat. Once you have experienced the delights you will never regret your decision.

Yours truly,

A cat lover

It's hard to state succinctly quite how great it is to live with a cat; I suppose the best way to find out is to experience it.

If the decision has just been made to add a cat to your family it's merely the beginning of the process. It's now necessary to look at the 'what' and 'where from' bits. There are a number of options available to you at this point.

Kitten or adult cat?

A kitten is the ideal choice if you already have a cat in residence. Adult cats will accept a kitten more readily as it poses little threat to them at this stage. Kittens can be a bit boisterous with the more staid adults, so it's important to give the kitten other things to do apart from 'bouncing on the big cat'. Chapter 7 will go into greater detail about how to make a stress-free introduction.

Kittens are always delightful to have around but they can be a handful. They will run up your curtains, disappear up your chimney, chew your fingers and generally act like hooligans. They probably aren't the ideal choice for people out at work all day; they need small frequent meals and lots of company when they are little, so if you are in a full-time job and contemplating your first cat it's probably best to look for an older one. If you are working but adamant about acquiring a kitten, then breeders will sell you two to keep each other company. However, many rescue charities will not re-home a youngster to a house that is empty all day.

Kittenhood is an expensive time for the owner, including as it does full courses of vaccination, neutering surgery, microchipping and all the acquisitions that are necessary to provide the best for your first cat (see Chapter 3). Tragically, cats between the ages of seven months and two years are statistically more likely to be killed on the road than at any other age, so there is an added risk if you plan to allow your cat to have access outdoors.

If you decide that you want to 'rescue' a cat and adopt a second-hand adult with an established personality, then you have a wide choice of options. Adult cats in need of a new home can be obtained from friends, neighbours, rescue shelters, vet surgeries or even your own back garden. Some cats will find you and turn up one day demanding food and a warm bed. If you fancy something more exotic than your average moggie, you can contact the various breed clubs that have 'rescue' co-ordinators who re-home their own kind to worthy homes. Most of my cats (including my most recent acquisition, Mangus) were pre-owned and I don't for one moment regret taking them on. I have had nothing but pleasure from all of them, but I do accept that taking on a cat with a vague or unknown background can be a bit of a lottery.

If you do opt for an adult cat, then don't forget there are oldies to choose from too. People tend to overlook the older cat for obvious reasons, mainly that they may get illnesses relating to their advancing years, so cost a fair bit and then they die! While I accept that this is often true, there are many advantages to taking on older cats. They will be past the stage of running up your curtains, straying, hunting or fighting. They are usually more tuned in to people and very fond of a good cuddle. They also seem incredibly grateful for a warm bed, a haven in which they can live out their days. When one day my present two, Bink and Mangus, are no longer here I have a feeling I may just act as custodian for a series of unwanted geriatrics.

Pedigree or moggie?

Many people decide on a pedigree as the ideal choice because they believe that it gives them a better idea of what they are going to get. This is true to a certain extent, but the next chapter will cover this in more detail. I struggle with the concept of spending a fortune on a pedigree kitten when there are thousands of healthy cats put to sleep every year for want of a decent home. If you are looking for a cat of a particular shape or colour, then I expect you would choose a pedigree. Don't forget, however, that the breeders will need to have done a good job in the first few weeks to provide the right sort of early experiences to make their kittens good pets. Owners of pedigrees also tend to follow breeders' advice and keep their pets exclusively indoors, thereby automatically curtailing their lifestyle to one of confinement.

Domestic short- or longhairs (the correct term for 'moggies') are nature's pedigree and they can be as beautiful, loving and intelligent as any expensive breed. They tend to be more robust in their health and probably more 'cat-like' than their pedigree cousins. It's your choice really.

Longhaired or shorthaired?

Longhaired cats are potentially high-maintenance pets, particularly the pedigree ones with a long coat that they cannot keep tidy themselves through grooming with their tongue. Most longhaired moggies have coats that are self-maintained; they may get a little knotting under the armpits but most of the coat will stay mat-free. Your cat would still benefit from a bit of additional grooming and this can be very relaxing for human and cat alike. If, though, you are considering acquiring a Persian, for example, remember that this beauty requires daily grooming. Just to make matters even more complicated, Persians often don't much like being groomed and they can put up quite a fight for such a normally placid breed. Despite their opposition it really is essential to keep on top of the combing as, if you miss a day or two, knots will develop that are virtually impossible to untangle without a visit to your vet or a professional grooming salon. You then end up with a cat sporting the latest poodle or lion cut and an expression that could curdle milk.

Shorthaired cats are less prone to the formation of hairballs in their stomachs (from grooming and ingesting fur) and are perfectly able to maintain their coat condition with their specially designed barbed tongue. Usually, though, this doesn't mean they don't enjoy a little brush/massage after a busy day in next-door's garden.

Male or female?

There are some owners who favour one sex over the other. They will tell you, for example, that male cats are more affectionate than females. From my personal experience I beg to differ: I have had six females and four males in my cat-owning lifetime and they all had their own unique personality; some were more affectionate than others, but certainly there was no bias in favour of the males. I believe that the cat's individual personality is more important than its gender.

There are some financial and practical considerations if you are acquiring either sex as a kitten. Males are cheaper to neuter and the surgery requires a shorter recovery time. (Ladies, be warned on this point: some men have a problem with the whole concept of castration. Do not let them influence you about the necessity for surgery at the appropriate time; it is the responsible thing to do.) Once castrated, male cats can develop into mummy's boys if their testosterone was the only thing that was ever going to make them brave. Owners rarely see this as a big disadvantage, although it can leave these cats susceptible to bullying outside. Females take the perceived trauma of an ovario-hysterectomy at the age of five or six months in their stride, with no obvious personality change seen in most individuals.

If you already have a cat of one sex it may be advisable to get one of the other, providing their personalities are compatible (for example, don't get an extra-bold kitten if your resident cat is shy and retiring). If you are acquiring two kittens, then one of each sex from the same family is probably the ideal combination; otherwise, pick the personality that suits you and worry about the workings at the back end later!

I hope this hasn't confused you too much and you now have

a clearer idea of what age and type of cat you are going to choose. Don't, however, get too locked into the idea that you 'must have a three-year-old ginger male shorthaired cat' when you go looking. It's like any shopping expedition – if you go in search of something too specific you will end up disappointed. By all means have a rough idea of what you are after and then be prepared to be flexible.

Getting a kitten from a domestic home

Some pet owners believe, erroneously in my opinion, that female cats should have one litter before they are spayed (neutered). Others like their children to experience the wonders of birth and the joy of little kittens. Unfortunately, another group of owners are either a little late in remembering to neuter their cat or don't wish to go to the expense and end up with a litter of kittens whether they like it or not. Inevitably all of these owners will advertise their kittens for sale at some stage.

The best places to look for 'kittens for sale' ads are:

- local newspaper
- noticeboard in your local shop
- noticeboard in your veterinary practice
- pet shop

Many kittens can be sourced via word of mouth; everyone seems to hear of a friend's neighbour's aunt with a litter somewhere in the country. There is a definite season for naturally produced kittens (pedigrees can be available all year round). Female cats are seasonally polyoestrous, which basically means that they will come into season at regular intervals between the months of February and September. After a nine-week gestation period they produce kittens during the warmer months, thereby allowing them a greater chance of survival. If you search for kittens during the summer months you are more likely to have a reasonable choice. In my experience those kittens available during the first four and last three months of the year tend not to be so robust, but I can't say I've ever seen any research findings to support that observation.

Some of the kittens advertised may be 'free to a good home'; others will be sold for anything between £5 and £100. I strongly recommend that you don't get tempted to pay a three-figure sum for a non-pedigree kitten. This may just tempt the individual to breed again from the mother cat rather than take the more responsible step of spaying.

Once you have identified the advertisement that appeals, the next step is to arrange a visit potentially to choose your new companion. Picking a kitten from a litter should always be a labour of the mind rather than the heart, but this is easier said than done. Selecting a kitten because you feel sorry for it or want to rescue it from a filthy, uncaring home is a noble act but it could mean long periods of distress and worry and a greatly reduced bank balance. It is certainly possible that you could have a long and enjoyable relationship with a kitten acquired in this way. However, if you want to reduce the risks of chronic illness or personality problems, I would suggest you consider the following.

Research the litter first by asking questions over the telephone before you view. If the answers to any of the following questions raise doubts in your mind, it may be better to look for another litter.

Question: Have the kittens been reared in your house?

The ideal answer to this question would be 'yes'. It's important that kittens are born and brought up in a normal domestic environment with all the usual sights, sounds and smells they are likely to experience when they leave to go to their new home. I always say that the more chaotic the home the better! Kittens that are born outside then 'grabbed' and brought into the house when they are four or five weeks old will not be well socialized and they'll always remain wary of people.

Question: Have the kittens been well handled?
Kittens between the ages of two and seven weeks are at their most receptive to learning about social relationships with other species. Research shows that handling by four or more people (men, women, children) gives the kittens the best possible start in life. If they are not being handled very much at all, this could be a problem, as they may never adjust completely to forming relationships with humans.

Question: Can the mother be viewed with the kittens?
If the mother is shy, nervous or unlikely to be around when people visit, then this reaction to human company may have influenced the kittens. I wouldn't necessarily rule out the litter based on this one negative reply but be mindful that the kittens may be nervous too.

Question: Has a vet examined the kittens and have they been wormed and treated for fleas?
It would be great if the owner is well informed and has taken all the requisite steps to ensure that the kittens have had all the necessary parasite treatment. If this isn't the case, you will probably incur a significant vet bill as soon as you take possession of the kitten.

Question: At what age are you allowing your kittens to go to their new home?
Pedigree kittens tend to go to their new homes at thirteen weeks of age, after a complete initial vaccination course. Owners of non-pedigree litters won't be offering this to prospective purchasers, so eight weeks would be the ideal time for the kittens to leave their mother and embark on their new lives. The owners may want to get rid of the kittens sooner because they are eating them out of house and home and annoying them, but I would suggest that is probably too young.

When you make the decision to view the litter, look for a kitten that fulfils the following criteria:

- bright eyes with no discharge
- clean ears with no evidence of dark brown wax
- shiny coat and no pot-belly (this would indicate a worm burden)
- clean anus with no sign of diarrhoea
- alertness and interactivity with the environment
- playful behaviour with the other kittens in the litter
- keenness to approach visitors

Getting a cat or kitten from a rescue centre

Rescue centres come in all shapes and sizes. Cats Protection re-homes the largest number of cats in the UK. The RSPCA (Royal Society for the Prevention of Cruelty to Animals) and Cats Protection have large and small shelters, depending on whether they are funded by headquarters or through local donations and legacies. The Blue Cross, Battersea Dogs and Cats Home and Wood Green Animal Shelter are some of the better-known cat re-homing charities. In such large organizations staff training is available but the quality of care and knowledge will always depend on the individuals working at each establishment. Other smaller charities exist that do equally sterling work with a minute fraction of the funding. There are many volunteers doing this job and some will even have pens in their gardens or second-hand moggies living in their spare bedroom.

Cats end up at rescue centres for so many different reasons. They may have been found as strays or been the victims of a relationship break-up, an allergic child, an emigration or a 'no pets' policy in a rented property or a nursing home. Some cats may have been usurped as the favoured pet when a new puppy came along or their owner may have died, leaving them alone. Once the cats have been signed over to the charity they start their journey towards finding a new home. They have first to be examined by a veterinary surgeon before they are put forward for adoption. Most centres are extremely thorough with health checks and they have local vets who provide charitable assistance at reduced prices to ensure the less fortunate of the cat world also receive good veterinary care. If the cat's background is unknown, before it is put on display it will usually be quarantined to ensure there are no diseases being incubated. During the quarantine period, or after if appropriate, the centre will usually perform certain checks including some if not all of the following:

- scanning for a microchip if the cat has been found as a stray
- castration of male stray cats if necessary (females are more problematical as it is not obviously apparent whether the surgery has already been performed)
- blood tests for FeLV (Feline Leukaemia Virus) and FIV (Feline Immunodeficiency Virus) in high-risk cats –

i.e., sick cats and entire male cat strays
- worming treatment
- treatment for external parasites – e.g., fleas, mites etc.
- veterinary examination including any necessary dental treatment

If the cats have come directly from a home along with a vaccination certificate it may be possible to waive these precautions, but they will undoubtedly be given a health check anyway.

At this point they will be available for viewing and prospective owners like yourself can visit a cattery, re-homing centre or private house during the relevant opening hours to have a look at the various cats to see if any take your fancy.

While you are getting to know all the hopeful little contenders it's worth sparing a thought for those people who look after them during their time of need, as they do a very stressful and unglamorous job. I know because I worked for the RSPCA in the late eighties/early nineties, so I feel I have an understanding of the business from that side of the fence too. Every day these volunteers or low-paid workers have contact with the people who are giving up their cats for adoption, most of whom have a perfectly good reason for doing so, and the process is heartbreaking for all concerned. However, there are some the rescue-centre staff consider feckless and irresponsible with regard to the standard of care they give to their cats. These people, unlike you or me, don't want their pets any more or make unreasonable demands on the charity workers rather than deal with their own personal problems. It is very easy to fall out of love with humans when you work for rescue organizations and many do. Sadly this can spill over into their contact with the public in general and I often hear the complaint that rescue-centre staff can be abrupt and judgemental with prospective adopters and too critical of what constitutes a suitable home for their charges. It's simple for me to understand how they end up this way but I don't necessarily believe this attitude is very constructive. My advice to you is: stick with it; at the end of the day it's giving a home to an unwanted cat that matters. There are literally thousands of unwanted cats in centres all over the UK and they need people like you to care for them.

Whatever the reason for cats being in need of a new home, they are undoubtedly confused and distressed in rescue environments, kept in tiny cages surrounded by other cats and being poked and stared at by a random assortment of humans. It is a very difficult time for these cats, so it's essential you take this into consideration when you view them. The individual cat's response to the cage and you can be very revealing and enable you to see how he may cope with adversity in the future.

Identifying stress in a caged cat

Stress in confined cats has been comprehensively studied and various scoring systems have been created (originally by Sandra McCune in 1992 and latterly by Kessler and Turner in 1997) to determine the level of stress via direct observation. This is referred to as the 'Cat-Stress-Score' (CSS) and it's useful to have a basic understanding of it when viewing a prospective adoptee. The scoring system starts at '1', indicating that the cat is fully relaxed, and slides down an increasing scale of distress to '7', referring to the cat as 'terrorized'.

A '1'-score cat would look like this:

Body – laid out on the side or on the back
Belly – exposed, with slow breathing

Legs – fully extended
Tail – extended or loosely wrapped round the body
Head – laid on a surface with the chin up
Eyes – fully or half closed; may be blinking slowly
Pupils – normal according to the ambient light conditions
Ears – half back (normal position)
Whiskers – lateral (normal position)
Vocalization – none
Activity – sleeping or resting

The scale continues in a sliding pattern to describe various body postures. Basically, if a cat is fairly relaxed he will be stretched out yet responsive to your contact but not over-alert. As a cat grows more stressed his body becomes upright and tense, the belly is hidden, the legs bend, the tail twitches or curls under and vocalization increases in frequency and intensity. The truly fearful cat may even attempt to escape.

It's very distressing to see a terrorized '7'-scoring cat but this is Kessler and Turner's definition:

Body – crouched directly on top of all fours, shaking
Belly – not exposed, with fast breathing
Legs – bent
Tail – close to the body
Head – lower than the body, motionless
Eyes – fully open
Pupils – fully dilated
Ears – fully flattened back on the head
Whiskers – back
Vocalization – plaintive miaow, yowling, growling or silent

You can see the wide variety of stressed responses to confinement. It's hardly surprising really as the enclosure itself causes only a fraction of the stress for these cats. They are bombarded with the threatening scents of other cats, strange people and unfamiliar odours. Routine is often absent as the cage is approached randomly and mealtimes may vary as the carers cope with the demands of running a rescue shelter. The display of anxious behaviour at this time is not necessarily an indication of how they will behave when introduced to a loving home with few restrictions and new routines. Don't, however, believe that a '7'-score cat will automatically respond to your home with gratitude and transform into a laid-back pussycat. The degree of stress response from a caged cat gives a reasonable indicator how that individual may cope with challenges in the future, taking into consideration any specifically distressing prequel to its arrival in the shelter in the first place. Even if the fear or anxiety manifests itself in a rigid body and sweaty paws, you are still potentially adopting a cat that will struggle with any number of difficult situations. If the cat's response is aggression it further complicates matters.

Finding out about their past

Many owners just don't get much information about the past life of their new acquisition. It isn't always the most important factor in adopting a cat, particularly as some of the information comes second-hand or is a vague representation of the truth. When I worked in a rescue cattery I coordinated the handover of many unwanted pets and several things became apparent then, namely:

1 People sometimes don't tell you the true circumstances for handing over a loved pet as they feel it will sign their cat's death warrant.
2 People feel guilty so gild the truth so that you don't hate them.
3 Cats are always three years old.

With regard to the latter, I'm not sure whether these owners lose count once the cat grows up or they have no concept of time. It could just be that they think if you knew the cat was really ten it could spoil its chances of re-homing. It is notoriously difficult to age cats unless they are very young or very old, but I have seen many adopted cats that new owners have acquired at the age of three and I would swear they weren't a day under fifteen! So what I am saying is: if the cattery supervisor seems vague about the cat's background, then it's entirely possible they either don't know the details or are not completely positive that they are true.

There are several things that may ring alarm bells, so a little close scrutiny and interrogation is sometimes necessary. These points may be raised by the shelter staff during the course of a conversation or may even be written on a sign on the cage door.

Alarm bells!

'Not to be re-homed with small children'

This could indicate:

1 The cat has a history of aggression towards humans. Specifying 'small children' could mean that a small child has been attacked previously, small children could potentially provoke attack, or the rescue centre fear they would be vulnerable with such an unpredictable pet.

2 The cat is nervous and the centre may feel that small noisy children would terrify it.
3 The previous owners have indicated a specific problem that involves small children.

'Must be homed as a single cat'
This could indicate:
1 A history of inter-cat problems, either with other cats in the household or in the territory. The resulting stress may well have manifested itself in a behaviour problem that the previous owner found unacceptable or impossible to live with.
2 The cat may have a condition or illness that is potentially contagious or infectious to other cats.

'Must be homed to a quiet household with no other pets or children'
This could indicate:
1 This poor soul has been given up for adoption because it's scared of everything. The previous owners or the cattery staff were unable to establish what combination of factors freaked the cat out so chose to be cautious and cover all potential stressors.
2 The cattery staff assessed the cat as nervous and thought a quiet home was necessary.
3 The cat has a history of a behaviour problem.

'Must be homed as an indoor cat'
This could indicate:
1 A medical reason or a physical disability that would make access to outdoors potentially dangerous for the individual or for other cats.

2 The cat has always been an indoor cat (not a good enough reason in itself).
3 The cat is extremely nervous and the cattery staff realize that it will probably choose not to go outside.

'Needs to go to a farm, smallholding or a working stable yard'
This could indicate:
1 The cat is virtually feral.
2 The cat has a history of aggression to people.
3 The cat has a history of territorial aggression and can't live in a built-up area.

There are many other 'alarm bell' phrases but the ones detailed above are probably the most common. All are worthy of further investigation; if you are really lucky the staff are adept at getting the truth out of previous owners and they will be able to give you more information to help you make your decision. A history of a behaviour problem in itself is not the end of the world as many problems are environment specific and if, happily, you fail to reproduce a similar environment in the future the cat will probably be fine. My own cat, Mangus, was second-hand when I acquired her because she had an expensive habit of eating leather (piça – the consumption of a non-nutritious material, see Chapter 8). Once she was removed from her multi-cat home to a place where she was queen of all she surveyed, she never touched an inch of leather again. I would always recommend you be armed with as much background on your cat as possible, so that you are prepared for all eventualities. The last thing you want is to take a cat to yet another household that just isn't right for him.

Adopting a 'special needs' cat

It's very difficult to walk away from the cat with 'special needs' because there is always the worry 'What if I don't adopt him?' or 'Who would want such a cat?' I totally understand the quandary as I have even succumbed to this emotional decision myself and adopted a 'scaredy-cat'. However, over the years I have spoken to literally hundreds of cat owners who have adopted an animal under these circumstances and suffered the consequences. There is always a feeling that love conquers all and kindness will reap rewards no matter how emotionally damaged the cat may be. Sadly, love is rarely if ever enough and a lifetime of stress-related illness, traumatic vet visits and disappointment usually ensues.

If, though, you genuinely want nothing back from this cat (i.e., no cuddles and no emotional feedback) and you have incredible patience, then you may be the right owner for a nervous cat. Just don't expect undying gratitude or affection any time soon.

My advice with my sensible head on would be to choose a cat that, when observed in its cage, would have a score nearer '1' on the Cat-Stress-Score scale described above and then you are more likely to have an emotionally robust creature that can cope with life's ups and downs. Always ask the cattery assistant to remove the cat from the cage and if possible allow you to have time in an area away from the pen where you can interact under slightly more normal conditions. Some larger rescue centres have rooms for this purpose that are laid out like a sitting room at home to give prospective owners the chance to see the cat behaving in a domestic setting. Unfortunately not all centres are so well equipped and you may only get the chance to hold them in a cramped environment with the other cats looking on.

Adopting a 'disabled' or elderly cat

Occasionally prospective owners will actively seek out those cats that they perceive will be left on the shelf if they don't personally take action. These are the geriatric, blind, deaf or amputee cats that sometimes find themselves in rescue shelters. It is a noble thought to give a home to the less fortunate of the cat world but it is not without its drawbacks.

Adopting an elderly cat

Pleasures

- doesn't stray very far and is often content just to sit in the garden on a sunny day and watch you weeding
- loves a warm lap
- rarely if ever runs up your curtains or destroys your furnishings
- has a wisdom born of age that is extremely endearing
- tends to interact more with people and 'talk' more

Pitfalls

- is in his twilight years, so don't plan too far ahead
- will probably require an indoor litter tray for the rest of his life
- tends to cost more in veterinary fees, as this is often the time that illness or disease becomes apparent
- may have an occasional toilet indiscretion or episodes of staring into space as senility approaches
- can be quite needy and howl at night
- may require 'aids' for the elderly – e.g., footstools or steps (to assist the arthritic in getting to a favourite high place) or extra thick bedding

I would always recommend an elderly cat if you have the funds and the emotional resources as he will reward you tenfold!

Adopting a blind or deaf cat

Pleasures

- purely dependent on the individual cat; if he's been blind from birth, it is often hard to tell he has a sense missing

Pitfalls

- you will have to provide a secure garden or confine the cat indoors
- moving your furniture around can be confusing for the blind cat
- deaf cats can be a bit noisy

How a cat copes with being blind or deaf will depend on the individual. Cats have such an excellent ability to adapt that most manage perfectly well, particularly if they lose their sight or hearing gradually. It's always worth checking how the blind or deaf cat came to be given up for adoption, so it can be established whether the previous owner experienced a problem relating to the disability.

Adopting an amputee

Pleasures

- as many as you would ever experience with a cat with a full quota of limbs
- marvelling at their ability to hunt, shoot and fish with a leg missing

Pitfalls

- you have to watch their weight, as obesity could be a major problem
- arthritis would be particularly difficult if it affected the one weight-bearing leg

Whether a missing leg is a major disability is a matter of perception. Admittedly cats with legs missing have a pronounced limp when they are walking, and scratching on the side of the head if a back leg is missing may need some assistance from you. However, watch an amputee chase another cat or run up a tree and it's often impossible to tell that there is anything remotely wrong.

There are other disabilities that may inflict a cat, some of which might require constant care and medication. It is always worth discussing this with a veterinary surgeon so that you fully appreciate the practical, emotional and financial implications of what you are taking on.

Home checking

Generally when you visit an adoption centre, particularly one run by a large animal charity, and make it known that you wish to acquire a cat you will be subjected to a process called 'home-checking'. A volunteer attached to the centre will visit you in your home at a convenient time and assess your suitability as a potential cat owner. One of the most important factors that this individual will be checking is the location of your house. Most rescue centres will not re-home a cat into a house on a busy road unless the space is suitable for an indoor lifestyle. They will, of course, be more flexible if you can give assurances that you will secure your back garden. They will also be checking that you don't have fifty-seven cats already or a large number of boisterous children under the age of five. Each charity or rescue centre has its own rules and regulations about who is and who isn't a satisfactory candidate. Some criteria, in all honesty, are dreadfully discriminatory and really don't have any genuine worth when establishing whether or not the environment would be suitable for a cat. Most places, in my opinion, are infinitely better than a long period of incarceration in a cage, but it is very difficult to create standardized procedures when so much of the home-checker's job is about 'gut feelings' and personal opinions.

Until such time as a better system is created or a radical shake-up takes place, it is a process that needs to be carried out. The visit will involve a number of questions about your family and lifestyle and maybe even some about your previous experience of keeping cats or understanding of what is actually involved in their care. If you are working you will probably not be allowed to adopt kittens, but they will certainly let you take an adult or two. If you show that you have the financial and physical ability to look after a cat, then it is entirely likely you will pass with glowing colours.

The adoption procedure

Once you have made your choice and passed the home-checking, you will be given a date to collect your cat and the necessary paperwork will be completed. You will have to sign an adoption form to confirm that you will give the cat the necessary care and return it to the centre if you can no longer provide a home. You will also be required to give a donation to enable the charity to continue to maintain its shelter and look after the residents. You will also receive, if appropriate, confirmation of the cat's microchip implant number and a vaccination certificate.

Taking the cat home

Preparation is the key to a calm introduction, so you will already have prepared your home and obtained the necessary bits and pieces that every contented cat needs (see Chapter 3). Often the staff at the rescue centre will give advice about settling in and introductions to existing pets, a topic covered in more detail in Chapter 7. Most rescue centres will advise that you keep your new addition indoors for at least three weeks to ensure he becomes fully acclimatized to his new home and is less likely to panic and stray in search of somewhere familiar.

This will be yet another potential drama for a rescue cat, as a period of confinement often leaves him in a state of constant anxiety. In my experience, most cats, having spent what seems like a lifetime in a prison, approach their new situation in a state of absolute delight. This is particularly likely to be the case if the lucky cat is in charge of all he surveys as the single feline resident. You will of course witness the initial withdrawal under the bed or similar hiding place as he takes stock of his new circumstances. There is really no need for you to panic at this stage: the cat is best left there as you go about your business and allow him to decide alone when it is safe to explore. Don't feel compelled to squeeze yourself behind the sofa with handfuls of cooked chicken to make him feel at home. This, in his opinion, would not be the best start to your relationship.

Cats will occasionally in the first few days (or weeks, if they are particularly shy) only eat and use their litter tray in the dead of night. The odd really uptight soul may even fail to do either for the first twenty-four hours. Once again, don't panic, as this is a part of the process that is best ignored. Keep putting down fresh food and checking the litter tray for any obvious sign of action and let nature take its course. If the situation persists beyond this period, then it may be wise to consult your veterinary practice or the rescue centre from whence he came.

Hints and tips for the settling-in period

- Choose a room to confine your new cat while he adjusts to his new surroundings. This is particularly important for tiny kittens as they need to be near their litter tray at all times in the early weeks to make sure they get their toilet training well and truly sorted!
- The room should have no fireplace, loose floorboards or any other potential hazards that a frightened cat may attempt to use as a hiding place. With that in mind, the kitchen is not a sensible place due to the dangers of the cat squeezing behind electrical appliances and fitted cupboards.
- Place a litter tray in a corner of the chosen room, food bowl away from the tray, water bowl away from the food, and a bed and toys elsewhere.
- A thick fleecy bed is essential for young kittens; a source of heat such as a heat pad or microwave-heated bag would be helpful in the winter months provided it is used according to the manufacturer's instructions.
- Ensure the same rescue-centre food is fed initially, particularly to kittens, to avoid any bowel disturbances associated with a sudden change in diet.
- Maintain the same litter material as the rescue centre during the initial period.
- Play with your new cat or kitten if he seems amenable but don't persevere if he becomes anxious.
- Don't coax the cat or kitten out of a hiding place; spend time in the room, reading a book or just sitting there, without paying him any attention, so he gets used to your presence.

- Offer small amounts of tasty food by hand if he looks as though he wants to make friends but is a little shy.
- Use fishing-rod toys to encourage your new cat to play to break the ice between you.
- Always cease interaction if the cat looks frightened.
- Don't persevere with approaches if your cat becomes aggressive; he's probably just scared and confused and would prefer to meet you in his own time and on his own terms.
- Be patient, particularly if he seems reluctant to eat or come out from a hiding place.
- Maintain contact with the staff at the rescue centre for reassurance and guidance.

Acquiring a kitten from a pet shop

As I always preach that kittens should be observed and chosen in the environment where they were reared, it would be wrong of me to recommend that you purchase a kitten from a pet shop. Historically these kittens often are too young to have left their mother, are from an unknown background and kept in poor conditions. If everyone stopped purchasing from pet shops, then the trade would cease and we would have better control over the decision based on the ability to observe good husbandry in their place of birth. Even grand department stores that sell expensive pedigrees would have to go a long way to convince me that this is a satisfactory way to acquire a kitten. I think it would be a much better idea for pet shops to advertise litters of kittens if they wish and allow the prospective owners to contact the breeder direct.

Acquiring a feral kitten

If this is something that appeals to you, it's important you have the facts and understand the distinction between a feral cat and a stray. A feral cat is one born in the wild with no history of human handling. They can live solitarily or in a colony usually consisting of a familial group. Once these cats are mature they will be extremely wary of humans and often attack viciously if cornered. Occasionally, over time, certain individuals will learn to trust specific humans if they are consistent food providers in a predictable location. If you acquire their trust over years you may just be able to touch them when they are feeding.

There are an estimated one million feral cats in the UK. This population expands without the control of neutering that we wisely carry out on our pets. Although conditions are harsh for feral litters, just over half of those produced will reach sexual maturity and potentially contribute to the population themselves. If you do the maths, then a single female could be responsible for a staggering 3,200 offspring in her lifetime. These cats potentially spread disease to the domestic population, so you can see how important it is for all those unsung heroes at cat charities to do their work with the feral community.

There are two fundamental approaches to the care of feral cats, both of which have control of numbers at the forefront of their systems.

One policy is referred to as Trap-Neuter-Release (TNR); it concentrates on identifying feral colonies, trapping individuals by means of humane traps baited with food, neutering them and returning them to their familiar territory. This stabilizes the group numbers by preventing any further expansion and enables sick cats to be treated or painlessly put to sleep. Many if not all of these cats are infected with fleas and worms and may even be carriers of the two most dangerous cat viruses, Feline Leukaemia Virus (FeLV) and Feline Immunodeficiency Virus (FIV). Often dental disease and malnutrition are also an issue, with the females lacking sufficient good-quality food to keep themselves well nourished, so that kittens are born weak and unable to cope in such harsh environments. Unfortunately it is virtually impossible to catch all members of the group, but by constantly dipping in it is possible greatly to reduce their impact on the environment. Neutered cats are less likely to roam great distances and fights with neighbouring domestic pets will be less intense in the males if they're not fuelled by testosterone.

The second approach is also to trap and neuter but, instead of releasing them, to attempt to put kittens in some kind of 'rehab' to turn them into domestic pets, thereby saving them from the terrible fate that is 'living like a feral'. Weaned kittens may succumb to the allure of the food-baited trap but, often, pregnant females have to be caught and confined to give birth in a rescue shelter. This policy relies on the premise that a cat's natural state is one of domesticity and that only owned pet cats have quality of life. This can actually become a complex and almost philosophical debate; do we have the right to take a species that is not domesticated and attempt to turn it into a pet just because we think it will have a 'nicer' life? I would agree that living in the average cat owner's home is a very cushy number in comparison, but the recipient of all this restrictive luxury has the ability to adapt and benefit from all the potential pleasures through generations of domestication. The feral kitten is approaching this way of life as a first-generation candidate and there is absolutely no guarantee that he will find it anything less than terrifying.

Trapping a feral queen and confining it to await the arrival of kittens is fraught with danger and can often result in her cannibalizing the new arrivals due to the stress of giving birth in a cage surrounded by threatening humans. If this doesn't happen, then it is probably best for the kittens to stay with the mother, provided that her nutritional condition is good, to enable her offspring to benefit from her maternal antibodies to support their immune system. Unfortunately this makes handling of the kittens difficult, as mother is going to be very reluctant to let a scary human anywhere near. While all this is going on the kittens are learning from their mother's reaction and deciding for themselves that all humans are distinctly bad news.

Having said all that, good results can be achieved if feral kittens are hand-reared through necessity. It's really important that they are given the chance to mix with domestic, chilled cats to give them a better perspective on life with humans. If these kittens are handled during the sensitive period of development between two and seven weeks in a domestic setting, they stand a sporting chance of becoming cats that are able to cope with life as pets. Hand-rearing in a shelter environment is rarely satisfactory. Kittens have an innate fear response and 'hazard avoidance' strategies that kick in at about five weeks of age; if they don't get into a home situation before then everything becomes even more difficult. If a shelter is confronted with a feral kitten after it has been weaned, it can safely be said it is virtually impossible to tame it. If you have experience of doing just that (or know someone who boasts they have done so), I would suggest the kitten probably wasn't a true feral in the first place.

Should you decide to adopt a feral (after reading the above), then you will be taking on a cat whose personality is almost inevitably towards the 'scaredy-cat' end of the spectrum. A feral tends to be more reactive and vigilant than the average domestic (try moving your foot suddenly while sitting down and a kitten with feral origins will launch skywards) and often takes much longer to acclimatize to a new home environment. You may notice a big difference between how he behaved in his original domestic setting with the person who reared him and his new persona once you get him home. Ferals can become one-person cats, but if you are lucky you will acquire a kitten with a fundamental character that embraces sociability. These kittens will come round, but I should say, having assisted many owners with feral kittens over the years, that it takes a lot longer than you would expect of a normal pet cat. All the suggestions for introducing a new kitten to your home (see above, or Chapter 7 if you have existing pets) are relevant to feral kittens but you need a great deal of patience successfully to complete each step of the process.

Taking in a stray cat

A stray cat may look a bit ragged and may even not be neutered but it is a distinctly different animal from a feral. A stray is born and raised with people but is now, for whatever reasons, living rough and fending for himself. He is fighting for access to shelter and food: maybe catching rodents, breaking and entering cat flaps and stealing cat food, or scavenging from dustbins. A stray will maintain the capacity to form friendly bonds with people but become more suspicious and jumpy the longer he spends as a stray. If a stray is caught he will eventually relax and accept handling by new owners and become a pet again.

A stray can find himself in this situation for a number of reasons – for example:

- roaming too far from an established home and unable to navigate back
- left home owing to the presence of other pets, sporadic food provision or rough children
- bullied by other cats in old territory, therefore attempting to establish a new home range
- owners moved house and left cat behind
- escaped shortly after moving house and unable to navigate back
- unwanted pet, denied access to its home
- a need to 'move on' inherent in its personality

Unfortunately, due to the opportunistic nature of the cat, many well-meaning folk presume a perfectly rotund and glossy feline is a stray just because he looks up pitifully at the back door while pointing a paw repeatedly at his open mouth, demanding food. *This is not a stray.* This is 'Billy-Five-Dinners', who does the rounds every morning once his owner leaves for work and scoffs extra meals he really doesn't need. A stray's coat will be dull and probably his body thin. He will not be wearing a collar or smelling of perfume!

Strays often turn up at the home of the cat-mad lady in the village and there is a strong suspicion that word has got round to all the cats in the vicinity that the occupant is a soft touch. In reality the strong scent of other cats and the presence of a cat flap (and the food and shelter that is inevitably just the other side of it) are a strong enough draw for most hungry cats. The primitive urge to 'love therefore I feed' takes over most people (misguidedly, in my opinion) and they automatically feed a strange cat if he asks politely enough.

If, however, the cat is extremely thin, with its head disappearing into your dustbin or eating the bread you have put out for the birds, it is entirely possible that it is genuinely hungry and in need of sustenance. If you feed this cat he will return, on the basis that if you took pity on him once it is likely you will do so again. Soon you will be asking him into your home and he will be curled up on your lap with a genuine smile on his face. Before you get too attached, though, you owe it to his previous owner (and to him to a certain extent) to make some enquiries as to whence he came.

Steps to take when you find a stray

- Make a note of the cat's colour, sex and any distinguishing marks – e.g., one white sock on a black cat.
- Contact the local veterinary practices to establish whether they have a record of any missing cats answering his description.
- Register him with the local vets as a 'found' cat.
- Take him to your nearest vet and ask for him to be scanned for a microchip (a form of identification implanted under the skin on the back of the neck).
- If the cat is elderly, thin and frail do not presume illness, abandonment or lack of love. Often old senile cats wander off and forget where they live, so check with the neighbours-two-doors-down and you'll probably find some worried owners searching for their geriatric Tiddles.
- Put up posters that a cat has been found. Include your contact telephone number (not your name or address) or the local vet's if the practice agrees.
- Place advertisements in local shops, as above.
- Always retain some information regarding the cat's

distinguishing features to establish true ownership should anyone respond to your ad or poster.

If you are very lucky you may find a place at a local cat shelter where your new stray friend can stay awaiting the arrival of the missing owner. More likely, though, you will have to continue caring for the cat yourself. Rescue charities allow a 'reasonable' period to elapse – usually about ten days – for owners to claim their cats that have strayed. As I have said already, there will always be those occasions when the cat is merely attempting to solicit an extra dinner, despite having a perfectly good and loving home to go to when night falls. Most cats with this habit have previous convictions, so sensible owners have them microchipped or burdened with a huge tag that states DO NOT FEED THIS CAT!

If no owner is found (or one is discovered but no longer wants the cat), you are then in a position to decide whether or not the stray stays with you for good. It is probably wise never to start feeding an animal unless you have a pretty good idea it will stay with you or someone you know. Strays are often entire tomcats, so the first step should involve a trip to the vet's for castration. You will also need to de-flea and de-worm him and generally arrange for a thorough MOT, including blood tests, to avoid potentially introducing disease to your resident cats (if you have any).

I have to raise a cautionary note regarding the introduction of a tomcat to existing feline residents. Male cats can be extremely affectionate; they are often described as the ultimate lap cat. Owners automatically extrapolate this personality trait to the cat's reaction to other cats: 'If he loves me he's going to adore Sooty and Tigger!' Sadly, this is not always the case, even if, when he first arrives, he walks past the others politely and respectfully and adopts a quiet corner in a grotty room so as not to offend his new housemates. This will soon change. Once he has found his feet, got over any minor aches and pains and sussed out feeble Sooty and Tigger, he will launch an unprecedented attack and attempt to expel the unfortunate couple so he has you all to himself.

Stray cats enter your household in gradual stages anyway, so there is often no need to adopt the introduction strategies discussed later in this book. However, it would probably pay to read my suggestions and adopt any that are relevant to you to ensure the new addition is accepted as peacefully as possible by the existing cats.

Oscar's tale

Just before I leave the topic of strays for good, I would like to relate a story to illustrate how resourceful and wonderful cats can be. About fifteen years ago I knew a lovely lady and her family who lived in an idyllic rural setting in Cornwall. They were great animal lovers and their menagerie included dogs, cats, rabbits, ponies and numerous small furry creatures in tanks and cages. One day when I was visiting my friend and enjoying the sunshine while sitting in her beautiful garden, I suddenly saw out of the corner of my eye a rather strange wobbly beast of indeterminate colour emerging from a bush. I drew my friend's attention to it and we both watched with some surprise as it approached. It was a cat (of sorts); a thin, dirty, scraggy specimen that walked towards us with a definite list to one side but with great purpose.

We crouched down and encouraged its approach and were dismayed to see it collapse at our feet as it finally completed the Herculean task of crossing the garden. It suddenly started to have a seizure and as it lay on its side with its legs thrashing we looked on helplessly, presuming we were witnessing its demise. Moments passed and eventually all the violent movement ceased and the little cat was quiet. I reached down and was delighted to find it breathing but clearly exhausted. We rushed the cat, a male, to my vet's, where he was given fluid and warmth to get him strong enough to be examined properly.

The following day he was a much brighter but still very dirty, skinny cat and the vet said that apart from his being starving and ancient there was very little obviously wrong with him. My friend felt that fate had brought 'Oscar' to her door, so she took him home and gave him her own magic brand of TLC and he blossomed. The other cats and the dogs accepted his arrival with the sort of quiet resignation you often get in households with vast numbers of residents, and my friend was soon delighted to report that the dirty cat was actually tabby-and-white, with a grand total of two teeth and a permanent cheesy grin. Every time I went to visit he would jump on to my lap, wrap his paws round my neck and smother my face and neck with a strange saliva-based goo that only the most ardent cat-fan would have tolerated. I really didn't have the heart to reject his advances; he was so thrilled to be part of a caring family that he wanted everyone to share in his joy.

Oscar did, however, always maintain a link with his roaming lifestyle. He would disappear for hours on end and sometimes stay away for a couple of days before returning. He was often dirty and exhausted and he would sleep soundly before repeating the process twenty-four hours later. Sadly, many months after his arrival, Oscar went out on one of his constitutionals and didn't come back. We were all devastated and we searched for days afterwards, calling his name to no avail. Eventually we gave up and accepted the fact that he had probably come to the end of his natural life and taken himself off somewhere quiet to die; we didn't like to think he'd fallen victim to an accident or attack.

The house felt very empty without him and my friend missed him very much. This was not by any means the end of Oscar, though. Several days later he returned, but he wasn't alone. My friend watched him walk slowly towards the back door with an even more antiquated and decrepit cat close behind. Every few steps he would turn back and make a little chirrup and the doddery old thing following would squeak in return. She was as thin as Oscar when he first arrived and had a twisted and swollen mouth. A trip to the vet revealed that she had a nasty gum abscess; this was duly treated and the little female cat returned to her new adoptive home with Oscar. He was extremely gentle and sweet to her and hardly left her side while she was recovering. She soon became strong and fit and enjoyed her life to the full.

As soon as Oscar first arrived my friend had made enquiries in the village. She presumed that one so poorly did not have a loving home but she did the responsible thing. When the second cat came she relayed the story to friends and neighbours, joking that Oscar had brought his girlfriend to stay. One chance encounter in the Post Office revealed the truth: Oscar and Tabitha (the name given to the older cat) were actually mother and son, the last surviving members of a family of farm cats belonging to a dairy farmer in the nearby valley. The farmer had presumed they had both died of old age, as they hadn't been seen around the hay barn for some time. My friend was able to explain to the farmer what had happened and he was surprised that they had both made the long journey up the hill to her cottage. He clearly wasn't making any claim to ownership of such expensive creatures with no further use as rodent controllers, so he was glad to confirm to my friend that she was welcome to them!

Oscar lived for another six years before passing away peacefully after a short illness and Oscar's mum died a few months later. We always thought how proud she must have been of her resourceful son who had found such a lovely home when he was sick and then taken her there too when she was in need of help. Who said cats can't be altruistic?

CHAPTER 2

From Abyssinian to Turkish Van

BREEDERS AND PEDIGREES

IF YOU HAVE PREVIOUSLY HAD GOOD EXPERIENCE OF A particular pedigree or even taken a shine to a charming photograph in a magazine, you may decide to plump for a specific breed rather than a moggie. Pedigrees can of course be obtained from rescue centres or from the re-homing section of the breed society, but generally they are purchased as kittens. To a certain extent this allows you to shop around and get exactly the colour, sex and shape of cat that your heart desires.

If you have decided to purchase a pedigree it is important that you embark on this route with a clear understanding of all the potential pitfalls. It is very difficult (I have discovered while researching this chapter) to get honest, unbiased advice from the breeders themselves. Understandably they are in the business of 'hard sell', so any negative points about their progeny will not be so well reported. I have no vested interest in the cat-breeding industry, apart from a fundamental desire to do the best I can for the species I love so much.

What is a cat 'breeder'?

In theory, anyone can become a cat breeder. If, however, breeders wish to register their pedigree kittens they need to contact the Governing Council of the Cat Fancy or the Felis Britannica (the UK member organization for the Fédération Internationale Féline) and complete an application form to confirm that they have conformed to a number of specific requirements for the breed.

The Governing Council of the Cat Fancy was established in 1910 and is the largest organization that registers pedigree cats in the UK. It produces a General Code of Ethics for Breeders and Owners, which, basically, states both groups should look after their animals and provide veterinary care. It also dictates that breeders should 'not knowingly misrepresent the characteristics of the breed nor falsely advertise cats nor mislead any person regarding the health or quality of the cat'. They should also provide written proof of pedigree at the time of sale, written information regarding diet and general care for the new owner and offer to help with re-homing if circumstances require it.

There is no licence required to breed cats, so it is impossible to guarantee that all breeders are good breeders. In my experience there is a wide disparity in standards, with some individuals doing a great job and others exhibiting scandalous shortcomings. I have interviewed owners who have bought kittens from breeders established for many years, some of them even holding significant roles in their particular breed society, and I've been astonished to learn that there are those who take little if any care with early socialization of their kittens and that certain individuals are breeding in such numbers, in production-line outdoor pens, that it would be virtually impossible to give their progeny the best possible start in life. These breeders are also the ones who, despite paying lip service before the money changes hands, refuse to discuss problems that the new owners are having or accept any culpability for wool-eating, house-soiling, anxiety or any other problem that the poor little cat is displaying in adjusting to normal domestic life.

I have tried for years to educate breeders, as have many others, about their very significant role in the future behaviour of their kittens. They are not only responsible for the shape and colour of their cats. The far more important factor, in my opinion, is their responsibility to shape their charges' character and encourage their ability to embrace and enjoy life as pet cats. I do reach some breeders but the ones that listen are, by definition, the ones who want to do the best job possible. I preach to the converted all the time. By far the most successful way to influence breeders, I believe, is to educate and inform you, the future pedigree-cat owner.

While researching this chapter I asked my wonderful assistant, Clare, to visit a certain national cat show to speak to breeders. The report she produced on her return was very revealing although unsurprising.

1 The vast majority of breeders stated that there were no health problems associated with their breed.
  This simply isn't the case. The Feline Advisory Bureau recently published a list of inherited problems in pedigree

cats and this is available for owners and breeders alike. Every responsible breeder should be keen to find out about these potential problems and do everything in their power to eradicate them. Only one breeder during Clare's research was open about inheritable problems and her own endeavours to ensure her lines were clear.

2 All breeders were happy to sell their kittens into multi-cat households.
There was no attempt to qualify this statement. Rescue charities are now realizing the significance of 'one too many cats' in multi-cat households and are carefully advising prospective owners accordingly. Breeders should be aware of potential problems and be keener to place their kittens wisely to ensure their wellbeing.

3 All breeders questioned kept their cats indoors and recommended this to new owners.
I struggle with this on a very fundamental level. I cannot condone the production of kittens of any breed if their future freedom of movement is to be denied them. This is surely tantamount to breeding an animal in the knowledge that its lifestyle will be less than perfect.

4 The majority of breeders interviewed bred for looks alone.
This is a dangerous game as beauty is only skin-deep. The majority of cat owners expect their pedigrees to be good pets, particularly as many breeds are advertised as conforming to a personality type.

Further research showed that many people believe that buying a pedigree gives some sort of guarantee that it will have a particular character. Cat websites and chat rooms revealed some disturbing recurrent themes:

'Nowadays I want a cat that has a more predictable temperament and health background. I have had too many heartaches.'

'It's just that when you get a cat from a breeder you know exactly what you are getting.'

'If you have a purebred cat from a reputable breeder you are less likely to have health problems, you know what the attitude of the cat is going to be, and if you suffer from allergies or just don't like a layer of cat hair covering the furniture you can choose a hairless cat or at least one that doesn't shed as much.' (Hairless breeds are not hypoallergenic as it is a protein found in cat saliva, skin and sebaceous glands that many people are allergic to.)

'We paid the extra money as a guarantee that our kitten would be well socialized, healthy and loyal.'

Owners and breeders alike need to understand that not all breed members are the same. In all the information available it is impossible to find two breeders who describe their progeny in exactly the same way. However, they seem to have decided there are two common personality types of cat: 'intelligent and dog-like' or 'placid and laid-back'. Various anthropomorphic adjectives are usually dotted in between; these include: boisterous, outgoing, fun-loving, self-assured, confident, curious, talkative, people-orientated, loyal, gentle, cheeky, unflappable, etc. Many if not all of these attributes will be evident if the breeder puts as much effort into the upbringing of the kittens during the important early weeks as he or she does into the choosing of queen and stud cat. Both nature and nurture need to play their part.

I have listed below a quick A–Z glossary of all the more

popular breeds, together with a précis of the sort of information you will find on the relevant websites and literature. Also included, courtesy of the experts at the Feline Advisory Bureau, are the inheritable and congenital problems (those passed down in the genes or present at birth) that are known to affect the breed. I have detailed, too, the behaviour I have seen that is less than desirable in the various breeds. Please bear in mind as you read this that I see a relatively small number of pedigrees every year out of all the cats registered but the number is disproportionately large in comparison with good old moggies. It is fair to say that pedigrees are probably more likely to find their way to a pet-behaviour counsellor for a number of reasons, including:

1 the owner's high expectations of the cat to 'behave' in a certain way
2 the likelihood that these cats will be kept exclusively indoors – circumstances often associated with the development of behaviour problems
3 the potentially greater ability for owners to spend money on 'high investment' cats

It is also important to remember that many factors influence the development of behavioural problems, so describing them as traits of the breed could be considered somewhat misleading. I am merely reporting what I have seen.

There are many owners of pedigrees who would never have another breed of cat because they have a long history of owning adorable pets. The important thing to remember is that there are *no guarantees* that paying a lot of money for a cat will mean you get perfection!

AN A–Z OF CAT BREEDS

Abyssinian

What the breeders tell you . . .

Also known as Abys, 'Bunny Cat'
GCCF breed classification Foreign
Country of origin Abyssinia
Mrs Barrett-Leonard, the wife of an English Army officer, imported the first Abyssinian from Abyssinia (now Ethiopia) in 1868. The breed was first recognized in Britain in 1882 and was shown at the Crystal Palace in 1883. It is possible that very similar cats were brought to Britain from Southeast Asia and the Indian Ocean coast; genetic studies indicate that the modern Abyssinian is closely related to cats found in these areas. The early Abyssinians were crossed with British Shorthairs, and later with Oriental breeds.
Lifespan 9–15 years
Physical traits A medium-sized, shorthaired muscular cat, the Abyssinian has a ticked coat – i.e., each hair is marked with fine dark bands. It has a round, wedge-shaped head with tufted tips to its ears. Colours include the usual rich brown, blue, lilac and chocolate. The almond-shaped eyes are amber, hazel or green.
Character traits The Abyssinian is described as intelligent and curious, but this is tempered with a cautious streak. They are extremely loyal, becoming very attached to their families, even pining when they go away. The male will generally tolerate other cats well, but the female may prefer to be the only cat in the family. Both sexes can form strong attachments to dogs. Abyssinians thrive on human company and

are best placed in a household where people are usually at home. They need plenty of space, with opportunities to climb, so keeping them exclusively indoors is not advisable. They are not particularly vocal cats.

What the breeders may not tell you . . .

Reported inherited and congenital defects Progressive retinal atrophy (degenerative eye disease); pyruvate kinase deficiency (causes anaemia); congenital hypothyroidism (a form of dwarfism)

It is strongly suspected that other diseases including renal amyloidosis (a type of kidney disease) and myasthenia gravis (severe muscle weakness) are inherited.

Undesirable behaviour seen A tendency to over-groom associated with stress is reported but I've never seen this in my own behaviour practice. I would consider the Abyssinian to be one of the less problematical breeds in their behaviour.

Asians

What the breeders tell you . . .

GCCF breed classification Foreign

Country of origin UK

With the exception of the Burmilla (the product of an accidental mating between a male Chinchilla and a lilac female Burmese in 1981), the Asian cats were brought about by breeders creating Burmese types in alternative colours.

Lifespan Up to 15 years but can live well into their late teens.

Physical traits The Asian group is the name used for cats of the Burmese type with non-Burmese coat colour, pattern or length, namely: Asian Self, Burmilla, Asian Smoke, Asian Tabby and Tiffanie. They are described as elegant, muscular cats of medium size with a short wedge-shaped face. The coat of the shorthaired Asians is short and fine with a glossy appearance. The semi-longhair Tiffanie, described separately later, has a medium-length silky coat. They come in a wide variety of colours and patterns.

Character traits The Asian is described as alert, active and intelligent, with a very friendly disposition. They are like the Burmese in temperament but slightly less boisterous (not a bad thing!). They love attention but can be quite demanding, often following their owners around the house, crying for attention. The curiosity and friendliness of the Asian can often lead it to stray into visitors' cars or delivery vans, so some breeders advise that they are best confined to the house or a secure garden.

What the breeders may not tell you . . .

Reported inherited and congenital defects Thumbs up for the Asian as there is no record of specific inherited or congenital defects associated with this breed.

Undesirable behaviour seen I occasionally see Asians with regard to inter-cat aggression and urine spraying, but they certainly seem to be watered-down Burmese with all the positive and few of the negative traits.

Balinese

What the breeders tell you . . .

Also known as Some colours of an almost identical cat are called Javanese in the USA.

GCCF breed classification Semi-longhair

Country of origin USA

The Balinese first appeared in America in the 1940–50s, when longhaired kittens regularly appeared in Siamese litters. In 1973 a Balinese was imported into the UK and the breed achieved recognition and Championship status in 1986. Balinese breeding stock is still limited, so matings to Siamese are still encouraged to widen the gene pool.

Lifespan The average life expectancy is about 12 years.

Physical traits The Balinese is a longhaired Oriental type with the bone structure, coat colour and pattern of the Siamese (see later), a fine silky coat and bright blue eyes.

Character traits Playful, talkative, inquisitive and social, to name but a few! The Balinese is described as quieter than the more 'reactive' Siamese. Breeders state that these cats need company and if you are out at work all day, they advise you to acquire a cat pal for company.

What the breeders may not tell you . . .

Reported inherited and congenital defects Lysosomal storage disease (typically causing uncoordinated walking)

It is strongly suspected that other diseases including congenital strabismus (squint), small intestinal adenocarcinoma (bowel tumour) and lymphoma (cancer) are inherited.

Undesirable behaviour seen There are just too few Balinese around to be evaluated. I don't think I've seen more than a couple in my entire career. As they are so closely related to the Siamese, it is possible they may, if they do go wrong, follow a similar pattern (see below).

Bengal

What the breeders tell you . . .

GCCF breed classification Foreign

Country of origin USA

In the early 1980s an American geneticist, Dr Centerwall, bred the Asian Leopard cat with a domestic cat during the course of research into feline leukaemia. The goal in developing the domestic Bengal cat breed, apparently, was to preserve a strong physical resemblance to its wild ancestor yet produce a pleasant and trustworthy family companion. The first few generations, referred to as F1, F2 and F3, are certainly not pets in the true sense of the word despite the attempts of many unsuspecting owners to make them fit into a domestic household.

Lifespan 10–15 years

Physical traits The Bengal is a large muscular cat with hindquarters slightly higher than its shoulders. The coat is thick and soft with a spotted or marbled pattern.

Character traits Originally very shy (hardly surprising given their ancestry), but successive generations are becoming friendly and playful. They are extremely active and vocal cats (with quite a harsh cry) that love water.

What the breeders may not tell you . . .

These cats prefer the delights of the great outdoors to a life of confinement. This may not be the general advice of the breeders but, in my experience, these cats love to be outside.

Reported inherited and congenital defects None confirmed, but suspected conditions include distal neuropathy

Undesirable behaviour seen Where do I start? If you get a good one it is fantastic, but I see a disproportionate number of Bengals on referral for territorial aggression, inter-cat aggression, inappropriate urination or defecation and urine-spraying indoors.

Birman

What the breeders tell you . . .

Also known as The Sacred Cat of Burma
GCCF breed classification Semi-longhair
Country of origin Burma
The Birman cat is believed to have originated in Burma, where it was considered sacred, the companion cat of the Kittah priests. In 1919 a pair of Birman cats were clandestinely shipped from Burma to France and the breed established in Europe from their progeny. By the end of the Second World War, only two Birmans were left alive in Europe and a programme of out-crossing was necessary to re-establish the breed. The Birman was first brought into the UK from France in 1965.
Lifespan 12–16 years
Physical traits Birmans are large, stocky cats with semi-longhair silky coats. They have blue eyes and colourpoint markings like the Siamese. All four feet are white (referred to as gloves, socks and gauntlets).
Character traits Birmans are intelligent, gentle and quiet cats that make ideal family pets.

What the breeders may not tell you . . .

Some breeders recommend that Birmans should be kept

indoors, but they love being outside and rarely stray too far from home. They are not renowned for their hunting prowess.

Reported inherited and congenital defects Squint (crossed eyes); kinked tail; congenital hypotrichosis (lack of hair); corneal dermoids (skin on the surface of the eye); haemophilia B (clotting defect)

Undesirable behaviour seen I don't see many Birmans but those I do usually have inappropriate defecation or urination issues, occasionally related to chronic bowel problems. I can honestly say I see very few aggressive Birmans.

Bombay

What the breeders tell you . . .

Also known as Self Black Asian

GCCF breed classification Shorthair

Country of origin UK and USA

The breed as we know it today started in the early 1980s, another of the group of Burmese-type cats. Crossing the Burmese with American Shorthairs developed the American Bombay, but the UK version resulted partially from accidental mating of Burmese with non-pedigrees.

Lifespan 15 years plus

Physical traits The Bombay is a muscular cat with a glossy jet-black coat and copper-coloured eyes.

Character traits The Bombay demands attention and is known for its 'dog-like' sociability. It is reported to be even more vocal than its Burmese cousin with a temperament that 'sometimes resembles a delinquent teenager's' (I'm sorry, but since when has that been a good thing?). Bombays prefer cohabiting with dogs to living with other cats and they are described as being the most assertive in a multi-cat household. They are heat-seekers and love duvets and heat pads.

What the breeders may not tell you . . .

Reported inherited and congenital defects Curly hair and abnormally short tails have been reported.

Undesirable behaviour seen I've seen no more than a handful of Bombays for inter-cat problems in a multi-cat household and only the odd urine-sprayer.

British Shorthair

What the breeders tell you . . .

Also known as British, often abbreviated to BSH

GCCF breed classification: Shorthair

Country of origin UK

The British Shorthair is descended from the British domestic moggies that came to our shores with the Ancient Romans.

Lifespan 9–15 years

Physical traits The British Shorthair is a stocky cat, larger than the domestic moggie, with a short, dense coat and round face. The eye colour varies with the wide choice of coat colour.

Character traits British Shorthairs are described as friendly, laid-back and undemanding. They are claimed to be good with children and happy to stay indoors (although I personally think they love the delights of a garden).

What the breeders may not tell you . . .

There are quite a few contradictions in the description of the typical British Shorthair; in my experience, they can be quite aloof and often only affectionate on their own terms. I don't see many British for attention-seeking problems!

Reported inherited and congenital defects Polycystic kidney disease; haemophilia B and other bleeding disorders

It is strongly suspected that other diseases, including hypertrophic cardiomyopathy (HCM), are inherited.

Undesirable behaviour seen They tend to present with a variety of problems including dodgy toilet habits and low attachment to their owners!

Burmese

What the breeders tell you . . .

Also known as The Genghis Khan of the cat world (or is it just me that says that?)

GCCF breed classification Shorthair

Country of origin Burma

The origins of the breed date back to the 1930s, when a small brown cat called 'Wong Mau' was taken from Burma to America. In 1949 two Burmese cats were imported into the UK and in 1952 the breed gained official recognition with the GCCF.

Lifespan 15 years (18–20 is not uncommon)

Physical traits The Burmese is a strong, athletic cat with a close glossy coat and yellow/amber eyes. The breed comes in a wide variety of colours, the most traditional being brown, but including blue, chocolate, lilac, red and cream.

Character traits The Burmese is referred to as an outgoing and very energetic cat that likes human attention. Burmese often become very attached to their owners and are sometimes seen as more intelligent than other breeds. This is an extrovert breed and once you have had a good Burmese it's hard not to develop a passion for them.

What the breeders may not tell you . . .

If you keep the Burmese indoors, beware: the devil makes work for idle paws.

Reported inherited and congenital defects It is strongly suspected that diseases including hypokalaemic polymyopathy (muscle weakness), Feline Orofacial Pain Syndrome (form of neuralgia) and Type 2 diabetes are inherited.

Undesirable behaviour seen Where do I start? Wool-eating, cable-chewing, territorial aggression, urine-spraying and inter-cat aggression are probably the main issues for those Burmese that go wrong. They seem to be highly motivated to behave in a certain way and those that go round the neighbourhood beating up all the other cats rarely respond to behaviour therapy. I often remark, 'Beware the Burmese's second birthday,' as these are problems that arise after maturity. Siblings can fall out irreversibly and this can necessitate re-homing of one of the pair. Breeders often recommend that Burmese are kept indoors (that certainly would resolve any territorial-aggression problems), but in my experience they long to hunt, shoot and fish like any other regular cat.

Burmilla

What the breeders tell you . . .

GCCF breed classification Foreign

Country of origin UK

Burmillas originated from a chance mating between a Burmese and a Chinchilla in 1981.

Lifespan 15 years plus

Physical traits The Burmilla is a medium-build cat with a dense agouti coat (two colours on each hair shaft, giving a tipped effect) and a round Burmese-type face. Their eyes are green or amber.

Character traits The Burmilla is described by the breeders as demanding, mischievous, easygoing and fearless, amongst many other things.

What the breeders may not tell you . . .

Not a great deal to hide here, as far as I know!

Reported inherited and congenital defects Nothing specific is reported.

Undesirable behaviour seen I don't see many Burmillas but I do remember one that sprayed urine in the house anything up to 45 times a day! It was probably the exception rather than the rule, so don't be put off.

Chinchilla

What the breeders tell you . . .

GCCF breed classification Persian

Country of origin UK

The Chinchilla breed was developed from one female, called Chinnie, in the late 1800s to create the distinctive coat pattern that we see today.

Lifespan 10–16 years

Physical traits The Chinchilla looks like a petite Persian but it has a distinctive white coat with black tips, giving it a silvery sheen. Their eyes are green and they look as if they are wearing make-up as the eyelids are outlined in black.

Character traits They have a quiet, placid temperament. They are also described as independent.

What the breeders may not tell you . . .

Reported inherited and congenital defects Primary seborrhoea (skin condition); progressive retinal atrophy (causing blindness); lysosomal storage disease (causing a myriad different problems); congenital polycystic liver disease; polycystic kidney disease (PKD)

It is strongly suspected that other diseases, including epiphera (weeping eyes), prognathism (jaw malalignment), HCM (hypertrophic cardiomyopathy) and Brachycephalic Airway Syndrome (breathing problems), are inherited.

Undesirable behaviour seen They can have dodgy toilet habits.

Cornish Rex

What the breeders tell you . . .

GCCF breed classification Foreign

Country of origin England (Cornwall)

All Cornish Rex cats are descended from Kallibunker, a

wavy-haired kitten born to a straight-haired mother and unknown father on a Cornish farm in 1950. The early Cornish Rex kittens were stockier than the slim Rexes we see today. Pedigree British Shorthairs and Burmese were used for out-breeding. The curly coat is the result of a single natural recessive-gene mutation.

Lifespan 10–15 years

Physical traits The Cornish Rex has a distinctive wedge-shaped head and large ears, with a short, curly coat and whiskers. It is slightly built and has a long thin tail.

Character traits The Cornish Rex is described as an intelligent adventure-seeker that loves to be the centre of attention. It is not a lap cat, but prefers to play games with its owner.

What the breeders may not tell you . . .

Reported inherited and congenital defects Congenital hypotrichosis (reduced amount of hair)

It is strongly suspected that other problems, including umbilical hernia and hypertrophic cardiomyopathy (HCM), are inherited.

Undesirable behaviour seen I have seen only a handful of Cornish Rex in my career and I certainly wouldn't say that they showed specific traits.

Devon Rex

What the breeders tell you . . .

Also known as 'Monkey Cat'

GCCF breed classification Foreign

Country of origin England (Devon)

The Devon Rex breed stems from a curly-coated cat found in 1960, by Beryl Cox, that mated with a local female cat. She found the litter in a field at the end of her garden, and one of the resulting litter also had curly fur. Like the Cornish Rex's coat, the Devon's is caused by a recessive gene and inbreeding was necessary to perpetuate the breed, although the Devon gene is different. Within ten years the breed was recognized in Britain.

Lifespan 9–15 years

Physical traits The Devon Rex is medium-sized, with a small wedge-shaped head and large ears. The coat and whiskers are curly. I think the Devon Rex is a beautiful cat (eye of the beholder and all that), but many people consider my own example of the breed, Mangus, as being reminiscent of E.T. For anyone who hasn't seen the Steven Spielberg classic film, I should add that this is not a compliment.

Character traits Devon Rexes are renowned for their energy and playful behaviour. They are active climbers (apparently) and not lap cats. I say 'apparently' because my cat, Mangus, rarely climbs anything as she tends to fall off, so she sticks to my lap like glue at every opportunity.

What the breeders may not tell you . . .

Reported inherited and congenital defects Myopathy (muscle weakness); vitamin K-dependent coagulopathy (bleeding disorder); congenital hypotrichosis (reduced hair)

It is strongly suspected that other diseases such as Malassezia dermatitis (skin problems), hip dysplasia and hypertrophic cardiomyopathy (HCM) are inherited.

Undesirable behaviour seen I have seen in my career only a couple of Devon Rex with very non-specific problems.

Egyptian Mau

What the breeders tell you . . .

Also known as Mau ('cat' in Egyptian)
GCCF breed classification Foreign
Country of origin Egypt
The Mau is a natural breed derived from the modern street cats of Egypt. They were first seen in Europe at a cat show in Rome in the 1950s and from there were imported into America in 1957. The Egyptian Mau was introduced into the UK in 1998.
Lifespan Up to 15 years
Physical traits The Egyptian Mau has a short, glossy coat marked with spots. Its eyes are green and its body is medium-sized and muscular. Often this breed has a slightly 'worried' expression (as do some Bengals) that is quite endearing.
Character traits The Egyptian Mau is playful, friendly, confident and a great climber. Breeders recommend that these cats be kept indoors as their striking looks may tempt others to steal them, but I think it's sad that they are denied freedom to roam outdoors. An outside run or secure garden would be a good compromise.

What the breeders may not tell you . . .
Nothing at all to hide here!

Reported inherited and congenital defects None reported.
Undesirable behaviour seen None identified.

Exotic Shorthair

What the breeders tell you . . .

GCCF breed classification Persian

Country of origin USA

The Exotic Shorthair is a shorthaired Persian, first seen in the 1960s. The breed came to the UK in 1984.

Lifespan 9–15 years

Physical traits The Exotic Shorthair is a medium-sized stocky cat with a large head. The eyes are also large and round-set and the coat is dense.

Character traits The Exotic Shorthair is gentle, affectionate and generally good with children. They are quite happy, according to breeders, to be left at home on their own and therefore make ideal house cats.

What the breeders may not tell you . . .

Reported inherited and congenital defects Primary seborrhoea (skin condition); progressive retinal atrophy (causing blindness); lysosomal storage disease (causing a myriad different problems); congenital polycystic liver disease; polycystic kidney disease (PKD)

It is strongly suspected that other diseases, including epiphera (weeping eyes), prognathism (jaw malalignment), HCM (hypertrophic cardiomyopathy) and Brachycephalic Airway Syndrome (breathing problems), are inherited.

Undesirable behaviour seen None identified.

Japanese Bobtail

What the breeders tell you . . .

GCCF breed classification Not recognized
Country of origin Japan
The breed can be traced back to the eighth century but it was not seen outside Japan until the 1960s, when the shorthaired Japanese Bobtails were imported into America. They carried the longhaired gene, which soon turned up in litters of kittens. The first litter bred in the UK was born in 2002.
Lifespan 9–15 years
Physical traits The Japanese Bobtail is a medium-sized, muscular cat. Its legs are slender and long, and the hind legs are noticeably longer than the front legs. The tail, when carried normally, extends about two or three inches, although it can be straightened out to about four or five inches. The hair on the tail is generally thicker and longer than elsewhere and grows in all directions to create the effect of a pompom.
Character traits Japanese Bobtails are extremely friendly and intelligent; they get on with most other animals and love human company.

What the breeders may not tell you . . .

This isn't a very common breed and, as yet, there are no reported congenital defects (apart from the tail) and no undesirable behaviour traits.

Korat

What the breeders tell you . . .

Also known as The 'good luck cat of Thailand'
GCCF breed classification Foreign
Country of origin Thailand
In 1959 two Korats were imported into America from Thailand and the breed started to attract attention in the 1960s. The breed was recognized in the UK in the seventies.
Lifespan 15 years plus
Physical traits The Korat has a heart-shaped face, green eyes and a dark-grey (blue) coat. It is a medium-sized muscular breed.
Character traits Gentle, intelligent and loving, the Korat is a talkative cat (according to some) that loves to play games. Korats mix well with other cats but some breeders comment that they like 'the upper hand' implying that they can be assertive in a multi-cat household.

What the breeders may not tell you . . .

Reported inherited and congenital defects Lysosomal storage disease (USA and parts of Europe)
Undesirable behaviour seen None that I know of.

LaPerm

What the breeders tell you . . .

Also known as Dalles LaPerm (or the 'bad-hair-day cat')
GCCF breed classification Foreign

Country of origin USA
The LaPerm was developed in the 1980s from a colony of farm cats in Oregon. The first queen was imported into the UK in 2002. As the initial group was very small, out-crossing played a part in the breed's development – including the Ocicat, Tonkinese, Somali, Tiffanie, Abyssinian, Burmese and Asian (what a mix!).
Lifespan 12 years
Physical traits The LaPerm has a curly coat in ringlets or short wavy hair, depending on the variety. It has a wedge-shaped head and a broad muzzle.
Character traits LaPerms are affectionate, outgoing cats that love people. They are rarely vocal, except when they really want something.

What the breeders may not tell you . . .

Reported inherited and congenital defects None reported.
Undesirable behaviour seen I've only seen one family of LaPerms and they were delightful. This particular group were suffering from the stresses of a multi-cat lifestyle that would have upset even the sturdiest moggie. I would have to say, therefore, that no specific undesirable behaviour traits are reported in the breed!

Maine Coon

What the breeders tell you . . .

Also known as The 'Gentle Giant'
GCCF breed classification Semi-longhair
Country of origin USA

The first Maine Coon was imported into the UK in 1983, having originated in Maine in the 1850s. The breed developed thick coats to withstand the harsh winters in that part of the world.

Lifespan 9–15 years

Physical traits The Maine Coon is the largest breed of cat. Its semi-longhaired coat is thick, waterproof and requires little grooming. The ears are large, with tufts on the tips, and the eye colour varies with the coat pattern.

Character traits According to several breeders, this is the man's favourite when it comes to cats! They are affectionate and outgoing, with a need to explore the great outdoors. They are also referred to regularly as 'laid-back', with a quiet chirping voice.

What the breeders may not tell you . . .

Reported inherited and congenital defects Spinal muscular atrophy (muscle weakness and tremors); hypertrophic cardiomyopathy (HCM)

It is suspected that other diseases, including hip dysplasia and patella luxation (displaced kneecap), are inherited.

Undesirable behaviour seen Nothing particular springs to mind. I've seen the odd household with Maine Coons not getting on but this certainly couldn't be described as a breed predisposition.

Manx

What the breeders tell you . . .

Also known as 'Rabbit Cat'

GCCF breed classification British
Country of origin UK
The Manx is thought to have originated about 250–300 years ago on the Isle of Man. It is generally accepted, however, that the Manx is a result of a mutation of the domestic shorthair.
Lifespan On average, 10–12 years
Physical traits The Manx resembles a British Shorthair, apart from the tail! Their legs are short, the back ones slightly longer. There are several distinct types called rumpies, stumpies, stubbies or longies, depending on the presence or otherwise of any residual tail length.
Character traits The Manx is described as highly intelligent, gentle and another of those 'dog-like' breeds. They are renowned for their love of humans and chasing just about anything. The breeders state they are quieter than many cats, using a distinctive long, monotone grunt for a miaow.

What the breeders may not tell you . . .

Reported inherited and congenital defects Sacrocaudal dysgenesis (mutation of the spine); megacolon and constipation; rectal prolapse; spina bifida; congenital urinary tract defects; corneal dystrophy (cloudy eyes)
Undesirable behaviour seen None specific.

Norwegian Forest Cat

What the breeders tell you . . .

Also known as Skogkatter (if you're Norwegian)
GCCF breed classification Semi-longhair

Country of origin Norway

The breed is believed to originate from local cats in Scandinavia breeding with shorthaired cats brought in by the Vikings from Britain and longhaired cats brought by the Crusaders. The breed was first recognized in Norway in 1930.

Lifespan 12–15 years

Physical traits The Norwegian Forest Cat is a large, muscular, semi-longhaired breed with a woolly undercoat that has excellent waterproof qualities. The ears are large and tufted.

Character traits The Norwegian Forest Cat is relaxed and easygoing. It tends to be confident and sociable, enjoying the outdoor life.

What the breeders may not tell you . . .

Reported inherited and congenital defects Glycogenosis (glycogen storage disease type 4), causing muscle weakness leading to heart failure

It is suspected that HCM may also be inherited.

Undesirable behaviour seen None reported.

Ocicat

What the breeders tell you . . .

Also known as Oci

GCCF breed classification Foreign

Country of origin USA

The Ocicat is a relatively new breed of cat that resulted from an accidental mating in 1964 between a Siamese and a Siamese–Abyssinian cross. One of the resultant kittens had a distinctive spotty coat, hence the name Ocicat after the Ocelot, and the breed was born. The first Ocicats arrived in the UK in 1988.

Lifespan 15 years

Physical traits The Ocicat is a medium to large cat with a muscular body and a short, lustrous spotted coat.

Character traits The Ocicat is a lively and athletic cat that needs plenty of stimulation and activity. It enjoys company and is rarely if ever aggressive towards humans.

What the breeders may not tell you . . .

Reported inherited and congenital defects None identified.

Undesirable behaviour seen I've seen the occasional Ocicat regarding inter-cat problems, but in insufficient numbers to define this as a breed 'trait'.

Oriental Shorthair

What the breeders tell you . . .

Also known as Foreign White and Havana

GCCF breed classification Oriental

Country of origin Thailand

This is basically a Siamese with green eyes and a different coat colour. It was classified as a separate breed in the United States in the 1970s.

Lifespan 15 years plus

Physical traits An Oriental has a long and angular head, with a long, slender body and tail. Oriental Shorthairs are the 'size zero' models of the cat world. They have green eyes, apart from the Foreign White, which has blue.

Character traits The Oriental is described as a loyal, affectionate yet demanding cat. The expression 'the devil makes work for idle paws' was made for this breed, so it needs plenty of stimulation and entertainment to stay out of mischief. Orientals also talk a lot, with a harsh miaow that always sounds as if they are complaining! They are good retrievers.

What the breeders may not tell you . . .

Reported inherited and congenital defects Lysosomal storage diseases (causing various problems)

It is strongly suspected that other diseases, including congenital strabismus (squint), kinked tail, feline asthma, small intestinal adenocarcinoma (intestinal tumours), mast cell tumours, thymic lymphoma (cancer) and amyloidosis (causing liver dysfunction), are inherited.

Undesirable behaviour seen I see a lot of Oriental-type cats, but Siamese are one of the most popular breeds, so don't be too alarmed. The main problems I come across are urine-spraying (they may cuddle up together with their cat companions but there's something about cohabiting that upsets a particularly sensitive Siamese), wool-eating, hyper-aesthesia syndrome and fur-plucking.

Oriental Longhair

What the breeders tell you . . .

Also known as Formerly called the Angora, but renamed in 2003. Also known as Javanese and the Mandarin elsewhere in the world.

GCCF breed classification Semi-longhair

Country of Origin UK

Bred initially by mistake in the 1960s by the mating of a Sorrel Abyssinian with a Seal Point Siamese.

Lifespan Up to 15 years

Physical traits The Oriental Longhair is slender, with an angular head and a long silky coat.

Character traits The Oriental Longhair is a 'born show-off' that craves company and needs plenty of stimulation to avoid boredom.

What the breeders may not tell you . . .

Reported inherited and congenital defects Lysosomal storage diseases (causing various problems)

It is strongly suspected that other diseases, including congenital strabismus (squint), kinked tail, feline asthma, small intestinal adenocarcinoma (intestinal tumours), mast cell tumours, thymic lymphoma (cancer) and amyloidosis (causing liver dysfunction), are inherited.

Undesirable behaviour seen None specifically reported.

Persian

What the breeders tell you . . .

Also known as Himalayan for the Colourpoint Persian

GCCF breed classification Persian

Country of origin Turkey and Persia (Iran)

The Persian breed has been established for at least 100 years. The early Persians looked like longhaired domestic cats but the appearance has become more exaggerated over the years.

Lifespan 9–15 years

Physical traits Persians have a long, luxuriant but very high-maintenance coat. They are relatively thick-set, with short legs and round, tufted ears. The head is large in proportion to the body and the face is short, with a snub nose.

Character traits Persians are said to be gentle, undemanding, and at their happiest when they are just lying around. They are not good climbers or jumpers, so they are not renowned for their roaming activity. Many breeders, if not all, recommend that Persians be kept indoors owing to the unusually long coat but it's rather nice to see a happy Persian skipping in from the garden with several twigs and a slug stuck to its trousers.

What the breeders may not tell you . . .

Some Persians have such an extreme head shape that they have difficulty breathing and suffer permanently blocked tear ducts. This cannot be comfortable.

Reported inherited and congenital defects Primary seborrhoea (skin condition); progressive retinal atrophy (causing blindness); lysosomal storage disease (causing a myriad different problems); congenital polycystic liver disease; polycystic kidney disease (PKD)

It is strongly suspected that other diseases, including epiphera (weeping eyes), prognathism (jaw malalignment), HCM and Brachycephalic Airway Syndrome (breathing problems), are inherited.

Undesirable behaviour seen You don't really want a cat with a high-maintenance coat that despises being groomed. Unfortunately, brushing has to be done daily and many Persians simply hate it; this results in constant battles or severely matted coats. I also see Persians regarding house-soiling problems; they tend to favour shiny floors instead of litter trays and they need a lot of convincing to return to acceptable habits.

Ragdoll

What the breeders tell you . . .

Also known as Cherubim or Ragamuffin
GCCF breed classification Semi-longhair
Country of origin USA
Ann Baker bred the first Ragdoll in California in the 1960s; Birman and Burmese cats were subsequently used in the breeding programme. The Ragdoll was first introduced into the UK in the 1980s, finally becoming eligible for showing in 1990. The unique quality attributed to the Ragdoll to feel little pain and go limp when picked up is a myth!
Lifespan 10–15 years
Physical traits Ragdolls are big semi-longhaired cats with blue eyes. They can have colourpoints, white feet or distinctive facial markings in the bicolour variety.
Character traits Ragdolls are referred to as easygoing and relaxed cats, with a friendly disposition, that are easy to handle. Some breeders recommend that they be kept indoors as they are so non-aggressive they are unable to defend themselves. This isn't the case in my experience.

What the breeders may not tell you . . .

Reported inherited and congenital defects It is suspected that crossed eyes, a deformed tail and HCM may be inherited.

Undesirable behaviour seen One UK-based website associated with the Ragdoll breed admits re-homing many adult cats 'who are exhibiting stress behaviours, because they are so lonely'. I can't comment on that, as I have never seen a cat exhibiting stress because it is lonely. I would say, however, that I see Ragdolls mainly for soiling in the house and general litter-tray issues.

Russian Blue

What the breeders tell you . . .

Also known as Archangels
GCCF breed classification Foreign
Country of origin Russia
The Russian Blue is a natural breed and not the result of selective matings with other breeds. During the Second World War the breed declined and attempts were made to boost numbers by crossing with the British Blue and Siamese. This resulted in nothing more than a blue Siamese, but a concentrated effort by breeders in the UK and America in the late 1960s resulted in a return to the original type.
Lifespan Up to 15 years
Physical traits The Russian Blue has a short, plush, blue-grey coat and bright-green eyes. It's a medium-framed cat with long legs and tail.
Character traits The Russian Blue is described as quiet, sensitive and shy by nature. Breeders recommend that they be kept indoors, but once again I would say that access to the outdoors is always preferable.

What the breeders may not tell you . . .
No bad news here.

Reported inherited and congenital defects Nothing reported
Undesirable behaviour seen I've seen only a few Russian Blues and they exhibited a wide variety of problems, so nothing seems to be breed-specific.

Scottish Fold

What the breeders tell you . . .

GCCF breed classification Not recognized
Country of origin Scotland
The gene for folded ears (the physical characteristic of the breed) is believed to have been around for 150 years, but modern Scottish Folds originate from a kitten found in Perthshire in 1961.
Lifespan 9–15 years
Physical traits The ears of a Scottish Fold are small and tightly folded, forward and down. The earflap is folded over completely to cover the ear opening and the flap is stiff and cartilaginous. The Scottish Fold is of medium size with a solid compact body.
Character traits The Scottish Fold is considered to be a hardy cat with a sweet disposition. They can be timid.

What the breeders may not tell you . . .
Folded ears may harbour ear mites or could impair hearing.

Reported inherited and congenital defects Osteochondrodysplasis
It is suspected that other diseases, including lower-urinary-

tract problems and HCM, are inherited. All Scottish Folds suffer from variable degrees of painful degenerative joint disease – hence their not being recognized by the GCCF.

Selkirk Rex

What the breeders tell you . . .

GCCF breed classification British

Country of origin USA

The first Selkirk Rex was bred in 1988 from a Persian male and a young female cat with curly hair that had been taken into an animal shelter. The breed was recognized in America in 1992.

Lifespan 9–15 years

Physical traits The Selkirk's body is quite chunky, with a round, broad head and short, curly whiskers. The distinctive coat can be either semi-long and wavy or short with tighter curls. Breeders recommend that owners 'spritz the coat with water and then scrunch the curls' to keep Selkirks looking their best. Oh, for goodness' sake!

Character traits The Selkirk is described with all the usual adjectives: playful, affectionate, sweet, dog-like, etc.

What the breeders may not tell you . . .

Reported inherited and congenital defects Hypertrophic cardiomyopathy, anecdotally

Undesirable behaviour seen No idea; never seen one!

Siamese

What the breeders tell you . . .

GCCF breed classification Siamese

Country of origin Thailand

The Siamese originated in Thailand (formerly Siam) and has been in existence for hundreds of years. The breed was imported into Britain in the 1880s and arrived in America shortly afterwards. The original Siamese colour was the classic seal-brown points with a warm-cream-coloured body.

Lifespan Up to 15 years

Physical traits The modern Siamese has a long body and tail, with a fine, glossy coat. It is a muscular cat with slender legs and large ears. Its eyes are a vivid blue.

Character traits Siamese are intelligent, vocal, playful, dog-like and capable of forming intense attachments to their owners. Breeders do suggest that they are best kept with other Orientals. They love company, both human and feline. They are good retrievers and easy to train to walk on a harness.

What the breeders may not tell you . . .

Reported inherited and congenital defects Lysosomal storage diseases (causing various problems)

It is strongly suspected that other diseases, including congenital strabismus (squint), kinked tail, feline asthma, small intestinal adenocarcinoma (intestinal tumours), mast cell tumours, thymic lymphoma (cancer) and amyloidosis (causing liver dysfunction), are inherited.

Undesirable behaviour seen All that bonding with humans sometimes has a down side. Siamese can develop a form of separation anxiety and get themselves incredibly uptight, often to the point of illness, when their owners are away. They can also develop attention-seeking behaviours that could drive you mad. I see Siamese about wool-eating, fur plucking, urine-spraying (one-cat-too-many syndrome), hyperaesthesia and anxiety.

Siamese (old-fashioned)

Also known as Thai, Traditional or Applehead Siamese. There has been a movement over the past few years to go back to the old-style Siamese with a less extreme body and head shape. This is how they should be!

Siberian

What the breeders tell you . . .

Also known as Siberian Forest Cat
GCCF breed classification Semi-longhair
Country of origin Russia
The history of the Siberian seems to date back to the fourteenth century and is surrounded by myths and legends! Their coat is an indicator of the harsh environment from which they originate.
Lifespan 12–15 years
Physical traits The Siberian is a heavy, solid cat whose coat is dense and weatherproof, with a heavy undercoat. A non-scientific study conducted in the USA suggests that this breed is less likely to cause an allergic reaction in humans, but this is yet to be proved.
Character traits The Siberian is considered to be intelligent

and independent. They need lots of space as they are very active.

What the breeders may not tell you . . .

Reported inherited and congenital defects HCM, anecdotally
Undesirable behaviour seen I have never seen a Siberian, but I imagine that you would get plenty of undesirable behaviour traits if you tried to keep this cat indoors!

Singapura

What the breeders tell you . . .

Also known as 'Singapore Drain Cat', 'River Cat'
GCCF breed classification Foreign
Country of origin Singapore
Its ancestors were reported to have inhabited open drains in Singapore. This is officially the world's smallest breed of cat. The modern Singapura reputedly stems from four cats brought to America from Singapore in 1975, though there is a suspicion that the breed may have originated from a cross between Burmese and Abyssinians. The first Singapura was imported into Britain in 1989.
Lifespan 9–15 years
Physical traits The Singapura, weighing only 2½ to 3kg, is muscular and well-proportioned, with a short, silky agouti coat (like an Abyssinian's). Their large eyes can be hazel, yellow or green.
Character traits The Singapura is an undemanding cat but loving and good-natured. It is a sociable cat, with both humans and other cats. Singapuras have a natural curiosity

that makes breeders suggest they should be kept indoors; some even say that they don't particularly want to be outside. A nice secure garden would be the ideal compromise.

What the breeders may not tell you . . .

There is nothing suspicious on record.

Reported inherited and congenital defects None reported.

Undesirable behaviour seen I have never seen a Singapura, but as far as I am aware there are no undesirable behavioural traits in the breed.

Snowshoe

What the breeders tell you . . .

GCCF breed classification Foreign

Country of origin USA

The Snowshoe originated in North America in the 1960s, when a Siamese breeder found three kittens with white feet in a litter. They were subsequently bred with Bi-Colour American Shorthairs. The Snowshoe was introduced to the UK in the 1980s.

Lifespan 9–15 years

Physical traits The Snowshoe is medium to large in build, with long legs and a short and glossy coat. The face, legs and tail have darker markings and the eyes are blue.

Character traits The Snowshoe is intelligent and inquisitive, quickly adapting to the presence of children and other pets. Snowshoes are described as laid-back and easygoing, but as they require more attention than most it's recommended that they not be left for long periods. They are also good retrievers.

What the breeders may not tell you . . .

Reported inherited and congenital defects Neuromuscular disease in young kittens, anecdotally.
Undesirable behaviour seen None reported.

Somali

What the breeders tell you . . .

GCCF breed classification Semi-longhair
Country of origin USA
The Somali is the longhaired version of the Abyssinian. The longhair gene was introduced into the Abyssinian breed in the early 1900s, but the longhaired variety was not bred specifically until the 1960s. Somalis were introduced into the UK in 1981.
Lifespan 9–15 years
Physical traits The Somali should look like an Abyssinian but with a semi-longhaired coat. They have a slim yet muscular build, large ears and a bushy tail. Their eyes are amber, hazel or green.
Character traits The Somali is described as intelligent, playful and mischievous without being destructive! It is also very people-orientated.

What the breeders may not tell you . . .

Reported inherited and congenital defects Progressive retinal atrophy (degenerative eye disease); pyruvate kinase deficiency (causes anaemia); congenital hypothyroidism (a form of dwarfism)

It is strongly suspected that other diseases, including renal amyloidosis (a type of kidney disease) and myasthenia gravis (severe muscle weakness), are inherited.

Undesirable behaviour seen I have seen only a couple of Somalis throughout my career and no alarm bells go when I hear the name!

Sphynx

What the breeders tell you . . .

GCCF breed classification Foreign

Country of origin Canada

Hairless cats are recorded throughout history as being a spontaneous mutation. The breeding programme began in 1966 in Ontario and the breed was brought to the UK in the 1980s. They have been recognized by the GCCF since 2006, at which time they promised not to recognize any more hairless breeds.

Lifespan 9–15 years

Physical traits The Sphynx looks like a bald Rex, although it is covered with a very fine, down-like hair. It is particularly sensitive to cold and the harmful effects of the sun, so the breed is condemned to a severely curtailed lifestyle. Its skin is wrinkled in folds and it doesn't have any whiskers. The Sphynx produces an oily secretion that breeders recommend is removed by bathing (what a life!).

Character traits Sphynxes are described as highly active, vocal and intelligent. They are inquisitive and agile; all these attributes have to be satisfied in the confines of your home. However, having a partially hairless Devon Rex myself, I say let them outside under supervision – they'll love it.

What the breeders may not tell you . . .
I have a question about the Sphynx: have the breeders ever wondered what it must feel like to lick bare skin with a barbed tongue designed to maintain a fur coat?

Reported inherited and congenital defects Malassezia dermatitis and furunculosis; paronychia; HCM
Undesirable behaviour seen None reported.

Tiffanie

(Not to be confused with the Tiffany: an American breed, not recognized in the UK.)

What the breeders tell you . . .

GCCF breed classification Foreign
Country of origin UK
The Tiffanie is the only semi-longhaired member of the Asian group. It first made an appearance in the mid-1980s, as a consequence of the experimental breeding programme for the Burmilla in the UK.
Lifespan Up to 15 years
Physical traits The Tiffanie is a semi-longhaired cat with a muscular body and bushy tail. Its legs are long and slim in proportion to its body. Its eyes are yellow or green.
Character traits Tiffanies are described as intelligent, sociable and gentle. They are active and curious, so, once again, breeders recommend they are kept indoors or in secure gardens. They are very vocal and often cry for attention.

What the breeders may not tell you . . .

Reported inherited and congenital defects There are none specifically reported, although the Tiffanie may have the same inheritable defects as the rest of the Asian group.
Undesirable behaviour seen None reported.

Tonkinese

What the breeders tell you . . .

Also known as Once called Golden Siamese, they are also known as Tonks.
GCCF breed classification Foreign
Country of origin USA
The Tonkinese is the result of crossing a Siamese with a Burmese. They were originally developed in the 1950s.
Lifespan Up to 15 years
Physical traits The Tonkinese is a medium-sized muscular cat with a fine, soft coat. The eye colour can range from green to blue.
Character traits The Tonkinese is active, vocal, affectionate and intelligent. Breeders recommend they be kept with other cats.

What the breeders may not tell you . . .

Reported inherited and congenital defects Lysosomal storage diseases (causing various problems)
It is strongly suspected that other diseases, including congenital strabismus (squint), kinked tail, feline asthma, small intestinal adenocarcinoma (intestinal tumours), mast cell

tumours, thymic lymphoma (cancer) and amyloidosis (causing liver dysfunction), are inherited.

Undesirable behaviour seen They tend to inherit some of the traits of both the Burmese and the Siamese. I've seen them regarding inter-cat aggression, wool-eating and urine-spraying.

Turkish Angora

What the breeders tell you . . .

GCCF breed classification Not recognized

Country of origin Turkey

The Turkish Angora is a pure, natural breed of cat, originating probably from the Manul cat domesticated by the Tartars. It migrated eventually to Turkey, where it is regarded today with great reverence as one of their national treasures. In 1962 American servicemen rediscovered the Angora in the Ankara Zoo in Turkey, in a controlled breeding programme dating back forty-five years.

Lifespan Up to 15 years

Physical traits The Turkish Angora has a silky, medium-length coat. Its eyes can be amber, blue or odd (one of each colour).

Character traits Turkish Angoras are described as quick-witted, quick-moving, and sometimes quick-tempered cats. They are among the most intelligent of the cat breeds, allegedly.

What the breeders may not tell you . . .

No secrets here.

Reported inherited and congenital defects None reported.

Undesirable behaviour seen None reported.

Turkish Van

What the breeders tell you . . .

Also known as Turkish Swimming Cat, Kurdish Van, Turkish Vankedisi (the all-white version)
GCCF breed classification Semi-longhair
Country of origin Turkey
The Turkish Van is a naturally occurring cat from the rugged area around Lake Van. It has an affinity for water and enjoys swimming. The Turkish Van came to the UK in the 1950s.
Lifespan 9–15 years
Physical traits The Turkish Van cat has a medium-long white coat, with auburn marks on its head and bushy tail. Its eyes are amber.
Character traits Turkish Vans are described as loving and intelligent. They are great athletes and they can wreak havoc in your home in the name of entertainment. As they are such water lovers, owners are advised not to leave the toilet seat up!

What the breeders may not tell you . . .

Reported inherited and congenital defects Anecdotally, HCM
Undesirable behaviour seen Nothing seen.

Other Breeds

There are an ever-increasing number of breeds being developed as we speak, none of them recognized by the GCCF and some for an extremely good reason. There is absolutely no excuse for breeding cats that are mutations or have a congenital defect that could in some way affect their quality of life. We have enough cats already: there is no room for 'niche markets' when an animal's life is at stake.

These breeds include:

American Bobtail (very short tail)
American Curl (ears curl backwards)
American Shorthair (a little like the British version)
American Wirehair (like the shorthair but with a wirehair coat)
Chartreux (like a slight British Blue)
Chausie (a large African Jungle Cat hybrid)
Cymric (longhaired Manx)
Kohona (a completely hairless Sphynx; bred by delusional fools)
Munchkin (a sad and immoral mutation)
Nebelung (like a Russian Blue with a semi-longhaired coat)
Ojos Azules (Spanish for 'blue eyes')
Peterbald (hairless)
Pixie-Bob (most have extra toes)
Savannah (a hybrid between an African Serval and domestic cat – why?)
Serengeti (a cross between a Bengal and an Oriental Shorthair)
Seychellois (a white-with-coloured-bits version of the Siamese)
Toyger (a Bengal cross Maine Coon for those people who want a toy tiger)

And if you aren't bored already here are some more:

American Keuda, American Ringtail, Antipodean, Australian Spotted Mist Cat, Brazilian Shorthair, California Spangled, Chantilly Tiffany, Cherubim, Desert Lynx, Don Sphynx, Foldex, German Rex, Khao Manee, Kurilian Bobtail Shorthair, Jungala, Malayan Burmese, Mojave Spotted, Neva Masquerade, Snow Cat, Sokoke, Sterling, Suqutranese, Ussuri and York Chocolate.

Top Ten Most Popular Breeds
(taken from the Analysis of Breeds registered by the GCCF)

	1995		2006
1	Persian (9,589)	1	British Shorthair (6,353)
2	Siamese (4,918)	2	Siamese (3,768)
3	British Shorthair (4,217)	3	Bengal (2,883)
4	Burmese (3,587)	4	Persian (2,693)
5	Birman (2,183)	5	Burmese (2,352)
6	Oriental Shorthair (1,355)	6	Ragdoll (2,126)
7	Maine Coon (1,338)	7	Maine Coon (2,122)
8	Exotic Shorthair (658)	8	Birman (1,740)
9	Ragdoll (583)	9	Norwegian Forest (1,199)
10	Tonkinese (564)	10	Oriental Shorthair (1,002)

Now you are armed with information regarding the variety of pedigrees that you can purchase, here are a few tips that you will need to take your buying decision forward. You are in the driving seat now and in order to eradicate bad breeding practices the prospective owners need to vote with their wallets. You are more likely to get well-brought-up kittens and better 'after-sales service' if you go to a responsible, knowledgeable and informed breeder.

WHAT YOU SHOULD EXPECT FROM A BREEDER

Kittens that are:

- healthy
- friendly, relatively confident and sociable
- fully vaccinated
- flea- and worm-free
- bred by individuals with a thorough understanding of breeding and rearing kittens
- bred by individuals who continue to undertake 'continuing professional development' to improve their knowledge
- registered with the GCCF or similar organization
- purchased with a money-back guarantee
- sold with a proven pedigree in writing
- sold on the basis that help with re-homing is available if necessary
- sold with a promise of ongoing advice if necessary

Anything less than the above is not acceptable. Breeders have an immense responsibility to produce good pets and it should not be undertaken lightly. However, some don't appreciate or understand the implications of bad husbandry and the kittens they produce are less than ideal. Unfortunately, by the time prospective owners have arrived at the establishment it is often too late to walk away. Many caring cat-lovers will take one look at quivering and sickly kittens and hand over money just to 'rescue' them from their dreadful situation. The GCCF and just about every important source of information recommend that buyers reject sick-looking or nervous kittens. This is easier said than done once you are standing right next to them and worrying what their future holds without you.

There is one easy solution to this. Once you have made the

decision to purchase a pedigree kitten you must allow your head to rule, at least for a short period of time, while you perform an initial sifting process over the telephone. There will be no scared little eyes staring up at you and no stinky establishment to walk into. A great deal of information can be gleaned over the telephone if you know the right questions to ask.

Questions to ask a breeder

How many cats do you breed from and how many do you have in your establishment?
If the breeder replies that they sell Burmese, Tonkinese, Bengals, British Shorthairs and California Spangles (for example), you can probably presume that this is a large kitten factory. With the best will in the world it would be difficult to provide the ideal early socialization for all the kittens. The more cats the breeder has, the more likely they will be to have problems with infectious diseases; these will always attack in environments that are overcrowded.

How many litters do you breed a year?
This is probably another way to glean the same information as above.

Are your breeding queens (females) kept in the home or in pens?
Some breeders keep their queens in pens and bring the kittens indoors only for short periods. This isn't satisfactory; kittens learn from observing their mothers and if they are brought in on their own they don't have the opportunity to see their mothers responding positively to things. The ideal situation is to have the mothers in a domestic setting and the kittens seeing

the delights of a home environment the moment they open their eyes.

Are your kittens handled and, if so, from what age and by how many different people?
The reply you want to this question is that the kittens are handled from two weeks of age for between half and one hour a day and that four or more people are responsible for this, including children and adults of both sexes. If the breeder says, 'Of course my kittens aren't handled, because that would spread disease,' then you need to say 'thanks but no thanks!' The risk of contracting disease from handling is far less than the problems that the kittens will experience as adults if they aren't socialized properly.

Can you tell me your programme for socializing the kittens?
You want to hear about their system for handling (from two weeks onwards), but you also want to hear that they ensure the kittens experience the normal sounds, sights, smells and textures of a normal domestic home – including children screaming, music, washing machines, vacuum cleaners and car journeys.

Are your kittens litter-trained and, if so, to what litter substrate?
Ideally the kittens will have been trained to eliminate consistently in a variety of proprietary litter materials and not shredded newspaper, for example.

Is there any history of inappropriate elimination (house-soiling) or any other behavioural problem in your breeding stock or past kittens?
They will probably say 'No'; you just want to be sure they mean it.

Is there any history of congenital or inheritable disease in your kittens?
You are looking here for conditions such as hypertrophic cardiomyopathy or hip dysplasia. See the breed list for further examples. A good breeder will explain to you the steps he or she takes to eradicate problems in their breed lines.

Do you provide support to new owners after they have taken your kittens?
Ongoing support is something that breeders really need to incorporate into their service, in my opinion. If they don't know the answer to any of the queries owners throw at them, they have an excellent source of information in, for example, the Feline Advisory Bureau to enable them to research and report back with good information.

While you are firing all your questions it would be great if the breeder fired some back. It's reassuring to know that breeders want to screen prospective owners of their progeny; this shows they care beyond the money. If you have received satisfactory answers to your questions it would then be sensible to make an appointment and view the kittens. Even then, if you are not happy about anything you see when you arrive, be brave enough to walk away. Bad breeding practice can only be stopped if prospective owners like you use buyer's power and give your money to those who are doing a decent job. No breeders can ever guarantee that they will produce the perfect kitten, physically and mentally. A helpful and caring response to problems should they occur is what you are looking for.

Good breeders are out there. They are not necessarily the ones who have been doing it for the longest time or the ones who breed the most kittens. I have, during the course of my career, met many excellent breeders. These are conscientious, intelligent and well-informed cat lovers who do a great job and produce great kittens; armed with the above information it's now your job to find them!

CHAPTER 3

Getting It Right from the Start

CHOOSING THE RIGHT VET AND KIT

ONE OF THE MOST IMPORTANT CONSIDERATIONS, EVEN BEFORE the arrival of your shiny new cat, is choosing the right veterinary practice for you. This is a decision made with both your head and your heart. For some years now I have seen a disturbing trend in veterinary practices. The emphasis has shifted to a much more business- and profit-orientated environment. There is talk of 'sales per transaction' and 'dental targets'; every client who walks through the door is a potential sale of food, wormer, toys, flea control, insurance . . . The list is endless. Now, I don't dispute that it is highly likely your cat needs or would benefit from all of those things, but I would feel much more comfortable if I was sold the goodies by someone really trying to do the best for my cat rather than earning commission. Fortunately there are still practices out there that combine the best of both worlds, embracing new technology while retaining the friendly old-fashioned ways of the village vet. I want my cats to be looked after by someone who would carefully weigh up the advantages of invasive treatments or diagnostic tests against the disadvantages of the stress that they so often cause. I honestly don't believe that you have to do new, expensive and clever procedures just because you can. I cannot imagine, for example, ever allowing one of my cats to have a kidney transplant. I have huge ethical problems with the choice of donor cat and the concept of adopting the poor mite after the surgery. Surely we should aim to give our cats the best quality of life rather than get all competitive about how long we can keep them alive?

So, in your quest to find the ideal vet for you, I would recommend in the first instance that you talk to friends and neighbours in the area and see what they have to say about the vet they use for their own animals. You are probably looking for phrases such as 'lovely man who helped so much when Charlie was ill' rather than 'looks just like a fancy private hospital with all the mod cons'. I strongly recommend that, once you feel you have short-listed the possible candidates down to two or three, you go and have a look for yourself, even doing what I have done in the past and requesting an appointment to see the vet without your cat to discuss the possibility of registering in his or her practice.

Every visit to the vet is a potentially traumatic time for your cat, so you will need to assess the suitability of the particular practice for your own needs as well as your cat's. Bear in mind that a stressed owner means a stressed cat. If you find the experience of visiting your vet a joyous one, then it is possible that your cat will take it in his stride too.

Here are just a few basic tips for things to look out for and questions to ask, starting from the moment you approach the practice's front door up to the moment your cat is discharged after a stay in the hospital.

Location of premises

Not all veterinary practices are purpose-built and many are conversions of houses or shops. This doesn't make the latter group any better or worse as veterinary practices, it just means that you may struggle to park nearby. If you are travelling by bus, check out the local bus routes and proximity of the nearest stop.

Reception

In an ideal world all veterinary practices would have a cat-only part where owners can wait with their cats prior to the consultation. This should be an area with minimum human and animal traffic going past, with raised shelving for placing cat baskets to avoid the risk of your cat coming nose to nose with a Rottweiler. It isn't always possible for the smaller clinics to have cats' waiting areas, but it's worth asking if dogs and cats are mixed freely. Some practices are particularly sensitive to these issues (gold star for those that are) and they ask owners to wait in their car with their dogs until their appointment. This usually keeps everyone happy, as many dogs get extremely agitated at the vet's.

The ideal reception area may have soft lighting, pet and general-interest magazines and a noticeboard with important pet information and pet-related dates for your diary. In the really friendly people-orientated practices you will find pictures of animal patients looking happy and healthy and thank-you cards adorning a wall. When I was managing a veterinary practice some years ago I always thought it was a good sign if a client took the trouble to write and thank the staff for the very thing they are paid to do: looking after animals and their owners!

Reception staff

Gone are the days when all medical receptionists were scary post-menopausal dragons who shouted out your (or your pet's) symptoms for all to hear and then made you feel like a hypochondriac (or pathetic cat owner) for daring to visit the surgery. Receptionists should be friendly, both on the telephone and face to face, and make you feel welcome and reassured as soon as you enter the surgery. They should have a basic knowledge of veterinary procedures – for example, timing for neutering, flea control and worming – but most of all they should know where you went on holiday last year, how many children you have, the name of all your pets and loads more trivia besides. These people are worth their weight in gold, particularly if they become confidantes of owners worried about their pets.

When you are looking for a new veterinary practice, check out the response from receptionists. They are almost always busy but, let's hope, never too busy to welcome a new face.

Veterinary nurses

Nurses, both male and female, are the backbone of the veterinary practice. Their versatility knows no bounds as they leap from talking to clients to mopping the floor, monitoring anaesthesia, giving tablets to fractious cats, caring for hospitalized patients and carrying out dental procedures.

There are usually a number of levels of expertise reached by the nurses and nursing assistants within a practice. Veterinary nurses have completed rigorous training and passed exams, both practical and theoretical. In 2007 a new register was set up for qualified nurses to volunteer to include their names. This makes them personally accountable for their actions (previously, the veterinary surgeon would have been responsible for any claim of negligence or lack of care) and ensures they complete continuing professional development courses throughout their career. These nurses are referred to as Registered Veterinary Nurses; they may have additional qualifications such as a veterinary nursing degree or a diploma in surgical nursing, for example. These are, academically, the crème de la crème! There will also be veterinary nurses who qualified some years ago who have chosen not to register and student nurses, in either their first or second year of training (this will only be the case if the practice itself is an approved training centre for nurses).

You may also see nursing assistants, kennel assistants or nurses who have never taken exams but have been nursing for twenty or thirty years and are absolutely brilliant. I wouldn't hesitate to put my animal in the care of any one of these excellent people. The most important thing is that each person's role within the practice and level of expertise is clearly defined, through lapel badges or a 'mug shot' on a board in the reception. It's important to know that you are being given advice by a senior and qualified person rather than an enthusiastic and over-confident student.

Practice manager

Some practices are big enough to justify employing an additional member of staff to fulfil the role of practice manager.

This individual may be a veterinary nurse or have experience within veterinary practice along with additional management qualifications. These managers are very rarely front-of-house people, but they will be diligently orchestrating the finances and the running of the practice from an office at the back of the building. They do a great job as it frees the veterinary surgeons to concentrate on what they do best: caring for your animals.

Veterinary surgeons

These are highly qualified, dedicated people with many years of studying and training under their belt. They are technically more highly qualified and academically gifted than your doctor, so that's got to be a big plus point when you consider placing your furry treasures in their care. However, not all vets are made the same. Some of the most competent and talented vets I have known or worked with over the years have been terrible at communicating with people. Owners have been thoroughly dissatisfied with their treatment and manner, even though, technically, they have done a faultless job in diagnosing and treating the pet. This is a simple illustration of the point that being a good vet is all about being perceived to be a good vet and that is achieved by being personable and a good communicator. Now, if you don't mind whether your vet is abrupt or pleasant and you want a no-nonsense approach to detecting and curing disease, this won't really concern you. You will decide on a veterinary surgeon based on those criteria alone. However, if you want to feel an empathy with your vet and, more importantly, that your vet has an empathy with your cat, then meeting the vet for the first time is when you allow your heart to rule your head. Ask yourself three questions: Did I warm instantly to this person? Does this person truly love animals? Will this person treat my cat as if he were his/her own? If the answer to these questions is '*yes*', I think you may have found your vet.

Consulting room

Now that you have met the vet you can look around the consulting room to see whether the facilities are cat-friendly. If you are very lucky you will have happened upon an exclusively-cats clinic. Dedicated feline fanatics staff these practices with a special interest in feline medicine. The consulting room, theatre and hospitalization cages will all be geared to suit the cat patient. A typical feline practice will have:

- Longer consulting appointments. This is essential as a cat's health is so greatly influenced by its environment and social situation, so discussion is needed to work alongside the physical examination.
- A non-slip consulting table. There is nothing worse for your cat than being handled by a stranger while unable to get a sure footing.
- A Feliway Diffuser. This synthetic pheromone plug-in device is a useful addition to the consulting room because it gives your cat a sense of familiarity in new surroundings.
- No equipment that the cat can escape behind or climb. It's more comfortable for the cat to be allowed to explore the environment if he so wishes and a gap behind a cupboard could mean losing your cat for the morning.
- A secure window with a blind that can be lifted to distract the patient. It's far easier to manipulate and examine a cat if he is suitably preoccupied.
- Secure doors; otherwise, you could lose your cat for more than a morning.
- All breakable drug bottles and equipment stored in

cupboards. This is just in case your cat takes fright and attempts the 'wall of death' around the room.
- Fine scissors and quiet clippers. These are for preparing areas of skin for injections and other procedures.
- Cat-sized needles, stethoscope, etc. So much equipment comes from the medical supplies for large animals or humans, but the dedicated cat practice will have everything scaled down to suit your cat's size.

If the practice you are viewing is not exclusively feline, then you may still find that the consulting room is equipped with all the necessary goodies and that all the above provisions are made anyway. This really would be a good sign that they truly appreciate the special care that the average cat demands.

Hospitalization cages and operating theatres

Some vets, once they understand that you are keen to do the best for your cat, will embrace the concept of being 'interviewed' by a prospective client and proudly show you the parts of the practice that you rarely see. Any vet reluctant to show you the back-room facilities may have something to hide. The timing of these tours may be dictated by the patients undergoing treatment, so don't be surprised if the vet requests you come back at a time when there's not urgent patient care going on.

When you do view the premises you are looking for an ordered environment with a high level of cleanliness, particularly in the theatre. Ideally there would be separate recovery and kennel facilities for dogs and cats but, once again, limited space often makes this impossible, so don't automatically exclude these practices. My own vet has mixed kennels and I haven't found a finer level of patient care in many other surgeries I've seen over the years.

You may want at this stage to ask various questions, such as:

- What is the procedure when my cat comes into hospital?
- What is the level of after-surgery patient care I can expect from your practice?
- Is a nurse or vet with my cat if he stays overnight in hospital?
- What are your out-of-hours and emergency provisions for patients in your care?
- Do you provide any 'well animal' clinics or advice?
- What is the continuity of care? Can I see the same vet every time?
- Do you refer many cases to specialists?

The answers you want to hear are that the practice keeps in touch with you when your cat is in hospital and they give you clear instructions in writing about post-operative care when he is discharged. If patients are kept overnight in the practice you want to be reassured that there is a knowledgeable person on hand to monitor his progress through the night. The vet may tell you that they use an out-of-hours service from another practice or dedicated emergency clinic and that really is quite normal. I would rather have a sleep-refreshed vet attempting to diagnose my sick cat than one who had been up all night performing an emergency Caesarean.

Some practices offer nurse-run clinics. These aren't essential but they can be extremely useful in identifying possible problems and nipping them in the bud. These clinics offer dental care and obesity advice (just like Weight Watchers – see Chapter 9) and even pill-giving, flea treatments, suture removal and dematting of the more truculent Persians.

One of the biggest complaints that many owners have about

modern veterinary practice is the lack of continuity of care. Most practices seem to exist with locums or a high turnover of assistant vets. This, sadly, is often unavoidable but every now and then you find a practice, usually a smaller one, where you can see the same vet (or two vets) most of the time. I believe this is an enormous bonus because it gives you an opportunity to form a good relationship and allows the vet to follow progress of a patient throughout its treatment. Clinical history is always recorded each time you visit but it is no substitute for familiarity with the patient.

Finally, and this is a point that many cat owners raise with me, you have to understand that your vet (no matter how good) is probably a general practitioner. There may be times when your cat is in need of a referral to a specialist in a particular field – for example, orthopaedic (bones) surgery, dermatology (skin) or oncology (cancer). I get a number of phone calls from people who feel their cat would benefit from a referral but their vet continues to treat to the best of his or her ability. For some reason these owners are reluctant to ask, yet I believe this is an absolute right and the best vets are those who know when the time is right to refer. You are well within your rights to request a second opinion on a diagnosis or treatment, so don't be afraid to do so; good vets understand your concern for your cat and will happily cooperate with another vet, either one within the practice or someone from a different surgery.

Pet insurance

Insuring their pet against accident and illness has become an increasingly popular option for many owners. It is the rise in insured pets that has contributed greatly to the advances in veterinary medicine. Pioneering and complicated procedures using high-tech equipment are extremely expensive and were impossible for many vets to employ before the arrival of pet insurance. However, there needs to be a good balance between doing everything possible to improve the quality of life, cure disease or save life and doing everything possible because the animal is insured. There are some procedures and diagnostic tests that, in my opinion, are not absolutely necessary for the overall welfare of the animal. I'm sure there are many vets who would disagree but, particularly in the case of cats, I believe the stress of the procedure should be assessed and weighed against the advantages that the procedure bestows on the animal.

There is a plethora of insurance companies offering policies for your pet, including the company you insure your car with, the supermarket you shop at and the store where you buy your underwear. Your veterinary practice may well have had a particular insurance company that they actively promoted, but legislation that came into force in 2005 has made it necessary for veterinary practices to be directly authorized by the Financial Services Authority in order to give advice about pet insurance. Most practices have not gone down this particular road, so they may be unable to give you specific guidance.

Although it might be something you'd rather not do, I would strongly recommend you shop around for your pet insurance and read the small print. There are a number of different levels of cover from each company and you need to ensure that you are paying for the policy that suits you best. For example, there are policies that cost less per month, but have lower sums insured per condition and/or higher excesses that are deducted from each claim you make. This is only a sensible choice if you never claim! You will soon find that most illnesses and traumas fall slightly short of your excess and you end up paying the bills in full *and* paying for insurance. You need to find a good compromise between level of monthly premium and the agreed excess.

Here are a few important points:

- Speak to other pet owners and listen to their experience with the various insurers.
- Insure your cat as soon as possible; the younger he is, the cheaper the premium.
- Insurance does not cover routine or prophylactic treatment (e.g., some dental work, neutering and vaccinations).
- You cannot insure for a pre-existing illness, disability or problem, so expect these to be exclusions on the policy.
- Don't be tempted to chop and change insurers when the premiums go up – any illness that has befallen your cat will be excluded on the new policy, so the change may not be cost-effective.
- If you decide not to take out insurance, then make alternative saving arrangements.

Some practices offer a loyalty saving scheme whereby you set up a standing order to your practice and this covers you for discounted routine treatments and something towards the unexpected bills. If this isn't available to you through your practice, it's advisable to make a monthly payment into a high-interest account, purely for any unforeseen vet expenses. You should have already factored into your original budget all the routine costs for keeping a cat – for example, food, vaccinations, worming and flea treatment – and this should be achievable, all being well, without dipping into savings. However, very few of us have the ability to pay a big vet bill without some difficulty, so an emergency fund is usually essential if your pet is uninsured.

The worst decision to have to make about treatment for a sick pet is one based on finances; from my side, it was probably the most distressing part of working in a veterinary practice. Many owners had to forgo treatment for their pet because they couldn't afford it. We always tried very hard to get any assistance we could for them when I worked in various surgeries. The PDSA (People's Dispensary for Sick Animals) will provide treatment for animals belonging to owners on income support, and charities such as Cats Protection may be able to assist some individuals on a case-by-case basis. Some practices are very understanding and will allow stage payments, but they are of course businesses and an excellent veterinary practice could potentially be forced to cease trading as a result of unpaid debts and the resulting lack of cash flow. It's a heart-breaking dilemma.

Worming

One of the least attractive attributes of our feline friends is their ability to be intermediate and final hosts for various unpleasant internal parasites. This basically means that a number of organisms spend either part of their life cycle in your cat or end up there, fully grown, and gaining nourishment while nestling in his intestines or lying dormant in other tissue within the body.

There are two common worms that affect cats in the UK: *Toxocara cati* (roundworm) and *Dipylidium caninum* (tapeworm). Kittens can become infested with roundworm by drinking milk from their mother if she was not treated for worms during her pregnancy. Other routes of infestation include ingesting worm eggs in cat faeces or consuming prey. Signs that your kitten has roundworms are fairly self-evident; he may have a slightly pot-bellied appearance, but vomiting a cluster of thin curled-up cream-coloured worms or passing them via the other end is usually the give-away. Tapeworms are acquired by ingesting fleas and the signs are often not so obvious. There may be no symptoms at all or you may see rice-like segments round your cat's anus. Heavy infestations can cause bowel disturbances and you may also notice your cat 'scooting' his bottom on the carpet due to the irritation in that area.

Without going into any more detail, let me just say that the most important thing is to worm your cat regularly. Your new kitten should be wormed at very specific intervals during the first few months, even if the breeder has already done so. Your vet will guide you towards the appropriate product and timing, so don't be tempted to pick up something from the supermarket. Wormers come in various forms – tablet, liquid, injection and powder – so there is no excuse if your cat is difficult to medicate. Worming an adult every three to six months is usually sufficient, depending on his lifestyle. Your vet once again will be the person to turn to for the practice's recommended policy.

Vaccination

Vaccinations are one of the unsung heroes of modern pet care. Without them, the cat population would suffer terrible diseases that are highly contagious and could result in death. If cat owners are responsible and caring, vaccination is a must-have.

Vaccines available

There are a number of diseases that you can vaccinate your cat against:

- Feline panleukopenia virus (FPV)

This is also known as feline infectious enteritis or feline parvovirus. It has an acute onset and often results in death.

Occasionally there are few symptoms apart from high temperature, depression and dehydration, but sufferers can vomit blood or have severe watery diarrhoea. Kittens born to mothers infected with FPV may have cerebellar hypoplasia causing head tremors, balance problems and a very distinctive wobbly walk. There is no cure for this, but it doesn't progress and they can grow to have a perfectly reasonable quality of life.

- Feline herpesvirus

This causes flu-like symptoms but the initial episode is short. Thereafter, however, the cat becomes a carrier of the virus and, at times of stress, can continue to have upper respiratory tract infections (for example, sneezing, runny nose and eyes) throughout its life.

- Feline calicivirus

This also causes 'cat flu', distinguished often by the presence of ulcers on the tongue or in the mouth. Cats can become carriers once infected and there are many strains of calicivirus. Vaccination is available against the most serious strains.

- Feline leukaemia virus (FeLV)

This suppresses the immune system, making the sufferer more prone to infections leading to, for example, cat flu and diarrhoea. It also causes anaemia, leukaemia and lymphosarcomas (tumours). It is spread by saliva through mutual grooming, sharing of bowls or bites from infected cats. It can also be passed to kittens in the womb via the placenta and during mating.

- Chlamydophila felis

This is a bacterial infection causing conjunctivitis and upper respiratory tract disease, with young kittens being the most susceptible. Symptoms can be managed with antibiotics.

- Bordetella bronchiseptica

This is a bacterial infection thought to be responsible for 'kennel cough' in dogs. It causes coughing or pneumonia in severe cases. The vaccination is squirted into the nostrils to act directly on the respiratory passages.

- Rabies

This infection is not an endemic disease in the UK but vaccine is available for those cats travelling abroad.

It isn't always necessary to have your cat vaccinated against every possible disease, as some cats are not at risk from certain infections. The essential vaccine is the one that covers feline panleukopenia virus (feline enteritis), feline herpesvirus (cat flu) and feline calicivirus (cat flu).

The others are more relevant to specific groups. For example, the feline leukaemia virus vaccine is recommended for any multi-cat households and all cats with access outdoors. It certainly isn't necessary to give this vaccination to a single cat kept exclusively indoors. The Chlamydophila felis vaccine is appropriate for multi-cat households, particularly breeders' establishments where the infection is endemic. It may be useful to vaccinate against Bordetella bronchiseptica prior to boarding your cat at a kennels that also boards dogs or if your cat is considered to be at risk.

A primary vaccination course of two injections, three to four weeks apart, is given to kittens, starting at nine weeks of age, to cover cat flu and enteritis. If you adopt an adult cat with unknown or no vaccination history you should still request the primary course. A booster vaccination is then given one year later. If you intend to have your cat vaccinated against FeLV, as well as cat flu and enteritis, your vet may recommend that the two vaccine courses are administered at different times to allow a few weeks between them.

There is a great deal of debate currently regarding frequency of vaccination boosters thereafter. Some vets are responding to recent evidence suggesting that these vaccines are effective for at least three years. You can elect to follow this protocol but the vaccine manufacturers still recommend annual boosters, so you may need to sign a form to give 'informed consent' for this regime to be followed for your cat. Even if you vaccinate your cat every three years I would still recommend he has annual health checks; these can potentially spot dental problems, weight increases or decreases or the early signs of disease.

Problems associated with vaccinations

There has been a lot of publicity recently about injection-associated fibrosarcomas. These are invasive tumours that develop at the vaccination site resulting from initial inflammation that doesn't go down. They are particularly related to vaccines that include a substance called an adjuvant, such as the FeLV and rabies vaccines. The incidence in the UK is estimated to be 0.04 cases per 100,000 doses of vaccine – much lower than the incidence of FeLV (also a fatal disease). Veterinarians in the USA routinely vaccinate against FeLV and rabies in the right hind leg to give a better chance of complete removal of the tumour should one occur. The most common side effects of vaccination are mild, including a transient lethargy and loss of appetite. If your cat or kitten develops any symptoms after vaccination it's always worth reporting them to your vet to put your mind at rest. I would recommend that you don't let any possible side effects put you off having your cat appropriately vaccinated; it is by far the best option to give him protection against disease.

Flea control

Most of us are in denial about our precious cats even coming within three feet of a flea, but I'm afraid they probably do. May I dwell one moment on the flea life cycle? The subject may not be one we like to think about, but it does put a different twist on things for those owners who say 'I've never seen a flea on my cat!'

The flea lives for about two years, during which time it spends most of its days on the host animal (that's your cat or dog), feeding off its blood. It can survive away from the host for up to six months, awaiting the arrival of another animal to hop aboard. A female flea lays eggs at a rate of about fifty a day; these eggs fall off the animal's body into the environment, where they hatch into the larval stage. The larvae don't like light, so they wriggle under your sofa (for example) and develop into pupae. The adult flea breaks out of its cocoon when it feels the vibration of an animal (or you) close by. If you do the maths you will see that there is a lot of unpleasant activity potentially taking place out of sight of the naked eye.

Has my cat got fleas?

These are the signs that you need to look out for to establish whether or not your cat is playing host to fleas:

- When you comb your cat you find tiny black flecks that are shaped like commas. If you put these on a piece of wet paper they will turn into blood! This is flea excreta and that means the little monsters are lunching somewhere on your cat.
- Your cat scratches a lot and chews at his fur; blood-sucking fleas can cause great discomfort.
- You have itchy bites around your ankles; if you are particularly delicious the fleas will bite you too.

- Your cat has tapeworm (cats get this from ingesting fleas during grooming).

Flea products Most of us have had a problem with fleas in our home at some stage, but they are easily kept at bay if you use the modern flea treatments that you can obtain from your veterinary practice. Many of the pet shop or supermarket products, together with the 'alternative' remedies for fleas, are fairly ineffectual, so it's safer in the long run to use what your vet recommends.

There are sprays for the environment that arrest the development of the flea's life stages so that adults do not emerge. Your home should be sprayed every six months (or according to the manufacturer's instructions) after thorough vacuuming. It's always best to pay special attention to areas under furniture and curtains, the cat's bed, against the skirting boards and down the sides of cushions on chairs and sofas. Various chemicals are used in these preparations so always read the instructions carefully.

Your cat is best treated with one of the spot-on treatments that are applied to the skin at the back of his neck. Most of these will kill fleas on contact with the coat before they bite so they are definitely best for those cats with an allergy to flea saliva. This is probably the most common cause of over-grooming and itchy skin in the cat. Flea collars, sprays and powders are old hat now and certainly not the most effective form of flea control.

The danger of dog flea-control products One very important point before we move on from this rather unpleasant subject: *never* give your cat the flea treatment intended for dogs. Some spot-on applications for dogs sold in supermarkets and pet shops contain an insecticide called permethrin. This is one of a group of compounds called pyrethroids and they are highly toxic to cats. When I was in practice I had the unfortunate task of nursing a couple of cats with symptoms associated with toxicity after their owners had not read the packet and applied a dog flea treatment. One of them, tragically, died.

NEUTERING

The responsible thing to do is to neuter your cat at the time advised by your vet. This is usually between five and six months of age. Some rescue societies will neuter their kittens younger before they are re-homed to ensure that there is no risk of progeny in the future. Providing the kitten is healthy enough to cope with an anaesthetic, then there is no intrinsic problem with this earlier neutering. Long-term studies examining the effects of early neutering suggest that the technique is safe, with no increased incidence of physical or behavioural problems associated with it.

Castration

Male cats are castrated. This means that the testes are removed under general anaesthetic, thereby removing the sperm bank and preventing the production of the male hormone, testosterone. This renders the cat incapable of fathering kittens, less likely to stray and less aggressive territorially. It also makes it less smelly in the home and, if you have ever smelt tomcat pee, you would have your cat castrated for this reason alone. The surgery is routine and straightforward and your cat is in and out the same day. There are no stitches to remove post-operatively and a full recovery normally takes place within forty-eight hours.

Spaying (ovariohysterectomy)

Female cats are spayed. Under general anaesthetic the veterinary surgeon will remove your cat's ovaries and uterus. As well as preventing her from having kittens in the future, it will also stop your cat coming into season and attracting the unwanted attention of stray tomcats. Spaying is a more complicated procedure than castration as it involves an incision made through the flank or through the midline of the lower abdomen. Once the female equipment is removed, the muscles and then the skin are closed, leaving the external sutures to be taken out usually ten days later. It is more important for the female cat to rest after surgery and many vets recommend that she is kept indoors until the stitches are removed. It is possible to spay a cat in season, although not ideal; it is also possible to spay a cat in the early stages of pregnancy. Don't subscribe to the myth that all cats should have one litter for their benefit, as this just brings more little cats into an over-populated society.

CAT KIT

Now that we've covered all the basic health considerations for your new cat we need to ensure you have all the right gear at home. There is a bewildering array of cat paraphernalia on the market and it is very tempting to spend a fortune on items that you are led to believe your cat cannot live without. I will therefore try to take the guesswork out of this shopping spree and give you a few hints for getting the right bits without spending a fortune or wasting your hard-earned money.

Cat carrier or basket

I would recommend you go for the plastic-coated wire basket

with a lid opening rather than a side opening. This will always make it easier to get a reluctant cat inside at the appropriate time. These baskets are easy to keep clean and, if your cat prefers more privacy, you can always cover it with a towel during the journey. I would recommend you line the base with plastic, newspaper and then a towel as this should be sufficient to deal with any 'accidents'. If your cat is particularly prone to this travel-induced incontinence, then a piece of thermal synthetic sheepskin bedding (available from good pet shops and your vet) can be cut to fit the bottom of your basket. Urine will go through the bedding on to the newspaper, which avoids your cat arriving at his destination all wet and smelly.

It's now possible to get lightweight carriers that look like holdalls with fabric mesh at the front. One such is an alternative for the larger breeds as it might be easier to carry and can be folded and stored flat.

Food bowls

I would, if I had my way, banish the food bowl for all those cats fed on dry food. I love feeding biscuits in problem-solving receptacles (more of that in Chapter 4), but it's always useful to have food bowls to give you the option to provide variety to your cat's diet and include a little wet food. Ceramic food bowls are the most generally popular and probably the best bet. Plastic scratches easily and can give off a slight odour that your cat may not like; stainless steel is easy to clean, but if your cat wears a collar it can make feeding a little noisy as any attachments bang against the metal. Beyond that, the choice of bowl size or shape is dependent on personal taste, so I would only add that some cats find it difficult to eat tidily from a plate without definite sides to it. Don't forget to provide at least one bowl per cat in your household.

Water bowls

Your cat needs to drink plenty of water, particularly if he is on a dry diet. Most owners make the mistake of offering water beside the food bowl, thinking that it's nice for a cat to have a little drink with his meal. Water will be more readily accepted if it is well away from food, ideally in a different room. Water bowls can, again, be ceramic and most cats like a nice big bowl (however, there are those that prefer a glass!). Providing a rain-water receptacle in the garden is a must as there are those cats that would much rather have mucky rainwater than the stuff we drink out of our taps.

Litter trays

There are a huge variety of litter trays available and it can be confusing deciding which one to choose. Check out which type your new cat or kitten has used previously, as it wouldn't be a bad idea to stick with that. I prefer to offer the biggest tray that the allocated area can accommodate, bearing in mind that you start with a small kitten tray but soon progress to an adult variety when your cat grows up. I know I preach familiarity as being important, but this doesn't include hanging on to the minuscule kitten tray and expecting your adult cat to squeeze into it.

Covered trays are useful to provide a discreet toilet in an area you would rather not share with cat poo. However, I generally recommend removing its cat-flap entrance, which always strikes me as being one obstacle too many for a cat just about to relieve itself. The flap also gives the bully cat in your home a chance to bash his victim round the head repeatedly when he's going to the toilet by pushing the flap backwards and forwards with his paw. Covered trays give more ammunition to the bully as well as it's great sport for him to ambush a cat coming out of a toilet by sitting in wait on the top; so be vigilant and if there's any sign of this sort of thing going on I would change to an open tray. You can always camouflage the site with the strategic placement of pot plants if you don't want it to be the first thing guests see when they walk into your home.

From bitter experience I would be wary of polythene litter liners and litter deodorants. If you wish to use them to make the whole cat toilet maintenance slightly more bearable, I quite understand, but if your cat suddenly takes a dislike to his tray and develops a penchant for your carpet, then liners and fragranced additives are the first thing to discard to combat the problem.

Litter material

We now have the benefit of litters made from paper, wood, silica, corn and fuller's earth. Many of them are targeted at you, the purchaser, rather than your cat, the user. If your cat was let loose in the supermarket aisles I guarantee that he (or at least most of his friends) would choose the most expensive, heavy and inconvenient of them all. I will always be a champion for sand-like litter substrates because I feel they best replicate the material that your cat instinctively favours. If, though, your cat will use the lightweight, biodegradable recycled paper-based products, for example, then I say 'lucky old you!' These are much easier to dispose of and, providing you clean out the tray thoroughly, just as good as any of the more expensive kinds for keeping odours to a minimum.

One last note on litter: please don't be tempted by those products that say: 'Doesn't need cleaning out completely for three months!' This is nonsense. If you do leave the litter for three months, merely removing the faeces that accumulate, it cannot possibly make the tray remotely appealing for your cat. *You* may not be able to smell it by the end of the third month, but I'm sure your cat will be holding his breath anywhere in the vicinity.

Scratching posts

Scratching is an essential maintenance routine for your cat. It removes claw husks, exercises muscles and marks territory. If you don't provide the right sort of scratching posts in the correct locations, your furniture, wallpaper or carpets will undoubtedly suffer.

Scratching posts come in many different shapes and sizes. They may simply be upright wooden posts covered with thick sisal twine or carpet, or they may also include platforms, beds, hiding boxes and dangling toys for the more energetic individuals. Before you choose one for your cat you may want to consider these criteria:

- The post must be rigid; cats when they scratch need resistance to do the best job.
- The post should be tall enough for your cat to scratch at full stretch. If you buy one for your kitten, you will need to change it when he grows up.
- The post should offer opportunities for your cat to scratch both horizontal and vertical surfaces.
- If it is a tall modular scratching post with various platforms and bed attachments it must be stable. There is nothing worse than a tall scratching post that falls over when your cat launches himself at it at full speed.

You don't have to spend a fortune on a scratching post; some of the most popular ones are ingenious designs using only corrugated cardboard. If space is an issue in your home you can always purchase the flat-panel type that can be fixed to your wall at the appropriate height. When you first introduce the scratching post to your cat there is always a temptation to grab his paws and rub them up and down the post to give him the idea. Try to resist doing this as it's virtually guaranteed to make him scratch your sofa instead. I prefer allowing the cat to explore in his own good time.

The position of the scratching post is important too; it may be good for you if it's placed out of sight in a spare bedroom, but your cat will have no particular desire to use it there. I would recommend that you try this instead:

- Place it near a window or radiator in a room your cat particularly favours.
- Position at least one scratching post near your cat's bed for use when he wakes up and has his early-morning stretch.
- Sprinkle loose catnip over the base.
- Play a game with your cat with a fishing-rod toy round the post so that his claws catch in the material; this often promotes a scratching motion.
- If it is a tall modular post, put dry food on the top to encourage use.

You can of course be adventurous and construct one of your own; some of the best scratching centres I've seen have been home-made. Don't worry that you will inadvertently train your cat to scratch your carpet if you cover the post with a similar material. I certainly wouldn't use an off-cut of your living-room carpet, for example, but if you choose a piece of hard-wearing material recommended for heavy-traffic areas it will be more durable and probably remain the only target of your cat's claws.

Cat beds

You can save yourself a great deal of money here by accepting that your cat will probably adhere to the philosophy 'if it's good enough for my owner, it's good enough for me' and use your sofa and your bed for rest and relaxation. The more independent soul will enjoy a cat bed, but a cardboard box lined with a soft blanket and with the front cut out is as good as any expensive faux-suede contrivance. I tend to purchase a selection of washable cat-friendly blankets and my cat Mangus alternates between spare beds, sofas, chairs and my desk in the office on those favourite of days when she comes to work with me.

If you want to wean your cat on to a bed that isn't quite so in-the-thick of the human family you could do a lot worse than investing in one of two heat-based options. Radiator hammocks are designed to hook over the tops of most styles of radiator and they provide a raised bed (always preferable) with an almost fur-scorching source of heat. For safety's sake I wouldn't leave the radiator in question on the highest setting, although I should say I've never seen a cat with contact burns acquired under these circumstances. The second option is an electrical heated pad, pressure sensitive, that can be placed under a cat blanket to give the most luxurious level of heat twenty-four hours a day. Mangus is a bit of an expert on such things, being a Devon Rex and not over-endowed with fur. She definitely favours the heated pad and it's the only way I can guarantee that she sleeps where *I* want.

Of course, there is rarely a substitute for your lap.

Cat toys

Once again, you can spend a fortune on toys but there really is no need. Providing you keep your cat's toys hidden away in a box and bring them out randomly to maintain their novelty, your cat will continue to play with them for years. The original choice of toy of course is paramount. Here are my suggested favourites:

- Real-fur mice. Covered in rabbit fur, these are a food by-product, so they're not as non-PC as they sound. It is the most natural of toys and the one guaranteed to get even the most sedentary cat playing.
- Octopuss. This is a home-made toy filled with catnip. Full instructions can be found in one of my previous books, *Cat Detective*. It never fails to please and is relatively easy to make, even for someone like myself whose sewing experience consists of attaching buttons to trousers.
- Fishing-rod toy. One with a feather attached is probably the most popular.
- Rubbish. Balls of screwed-up paper, tin foil, corks, walnuts (unshelled), cardboard boxes and paper bags will all cost nothing but give hours of entertainment.
- Hair bands. Hair elastics are just the right shape and size for retrieval games.

And, of course, if your cat is one of the two-thirds of the domestic cat population that absolutely adore the addictive nature of the catmint herb, then anything filled with catnip is sure to please.

Collars

There are a number of valid reasons why you may wish to fit a collar to your cat:

- to attach the magnetic 'key' for an exclusive-entry cat flap
- to give a visual sign to neighbours that he is not a stray

- to attach a tag with contact details

There are also a number of invalid reasons:

- the latest Cartier diamanté collar is a must-have
- he looks so cute in that studded number
- it combats fleas

Flea collars are not as effective as the latest spot-on treatments and no collar should just be a fashion statement, but if you feel your cat needs to wear a collar for his own safety there are some important considerations.

Type of collar There are many different kinds of collars on the market. The most important feature is its ability to release your cat if he gets it caught on something. Many collars are fitted with an elastic section, but this is only effective if the elastic is large enough to stretch sufficiently for your cat to pull his head out of it and escape. In my opinion, the best design has a snap-open section that releases when the collar is under sufficient strain, i.e., a cat's bodyweight. It's always worth checking these collars, as some are more resistant than others. If you hang a bag of sugar or a kilo weight from it and it doesn't snap open, then I would suggest it's a little too stiff for complete safety. I would also avoid hanging bells and extra bits on the collar as claws can easily get caught in them. The magnets that operate cat flaps are notorious for collecting scrap metal as your cat goes about his business, so don't be surprised if he drags in the odd masonry nail or fork.

Fitting the collar Ill-fitting collars cause a very common injury seen by vets. It affects cats that have been straying for some time. During the course of their period away from home they have got their collars caught. In their attempts to break free the collar has only enabled them to get a leg through it. Therefore they free themselves but end up with the collar partially round their neck and under one leg. This causes a terrible friction injury to their axilla (armpit) as it wears the skin away, which becomes infected. This is a notoriously slow wound to heal, as there is so much movement of the skin in this area. Fitting the collar correctly in the first place would avoid the risk.

A collar should be fitted surprisingly snugly; you should just be able to push two fingers through between the collar and your cat's neck. When you first fit the collar you will find that your cat may tense his neck muscles. You will push two fingers under, think you have fitted it correctly, and then when your cat relaxes it will look like a dangly necklace. Always re-check the fit after he's got used to the collar because you will probably have to adjust it. There is often a degree of protest when the collar is first introduced, including walking or running backwards, hanging the head down, scratching furiously and trying to grab it with his mouth. If you fit the collar just before mealtime or before his favourite game, it may distract him sufficiently to make the whole experience far less traumatic.

One final word on collars: if you want to take your cat for walkies, then don't attempt to fit a lead to a conventional collar. Harnesses that fit securely around the cat's chest and neck are the only appropriate device for this; they're safer for the cat and virtually Houdini-proof.

Microchipping

Now you have chosen your vet, bought the necessary cat accoutrements and rid your new pet of all nasty parasites, it is time to consider the implications of the great outdoors. The indoor-versus-outdoor debate is tackled at some length in Chapter 6, so if you are undecided you may want to read that section in some depth. If and when you do decide to give your precious new acquisition the freedom of the great outdoors, it is sensible to consider microchipping. This sounds a bit high-tech but it's an incredible boon for returning lost cats to their owners. A small microchip (the size of a grain of rice) is inserted by your vet, via a relatively pain-free procedure, under the skin between your cat's shoulder blades. This microchip has a unique number that can be read easily by a hand-held scanner used by the RSCPA, other rescue charities, the police, dog wardens and all veterinary practices. This number is then checked against a central computer where your contact details are registered. Hey presto, you and your cat are reunited.

One particular story about microchips springs to mind. Julie contacted me some years ago to say that, to her great relief, her lost cat Bob had been found after she had spent many months searching for him and he was now safely restored to the home. I was pleased for her but a little confused why she was telling me, until she said that Bob was struggling with being back. He had apparently become so aggressive with her that she'd had to shut him in the basement and wear thick boots and gloves when visiting to protect herself against the attacks he launched every time she tried to comfort him.

I duly went to Julie's home the following week and sat down with her (without Bob) to discuss the nature of his going and subsequent return. She had worked so hard to get him back, convinced that it should be possible as he was microchipped. It was quite a coincidence that someone in a village some miles away recognized the description of a stray she had been feeding and took him back home. Julie showed me photographs of her cat before he went missing: he was a rangy black cat with big eyes. I thought he was lovely as he reminded me of my old cat Bakewell, so I was keen to see what nine months on the streets had done to Bob to make him so angry.

He was still confined to the basement, so I decided to go down and investigate. I wasn't being particularly brave as it was quite apparent that his attacks occurred when Julie made desperate lunges towards him in a frustrated attempt to receive the grateful thanks she believed she richly deserved. I sat down on an upturned box and, sure enough, there, crouched in a corner, was a Bakewell lookalike with round, black eyes and flattened ears. As he got used to the idea that I wasn't going to grab him he relaxed and gently moved forward to investigate. Wow, it must have been a tough nine months! His chest was much broader, probably due to all that exercise, and the few white hairs that used to be on it were gone. Even more alarming was the fact that his legs were definitely shorter. I'd never seen that before.

I went back upstairs to talk to Julie and asked her casually if she could tell me which vet or organization had scanned Bob to read his microchip. Julie replied that it was just as well she had put her contact details on all the posters and mailshots because the wretched microchip had moved and migrated somewhere under Bob's skin and was therefore impossible to read.

The moral of this story is: it is rare for modern microchips to migrate; they usually stay put. In this case they couldn't find Bob's microchip because it wasn't Bob they were scanning.

CHAPTER 4

I Love, Therefore I Feed

FEEDING YOUR CAT

WE ARE BOMBARDED ON A DAILY BASIS WITH INFORMATION relating to cat food – why one product is better than another or why feeding a particular 'meaty chunk' will prolong your cat's life. It's hardly surprising that so much advertising takes place, as the cat-food market represented a staggering £708 million worth of sales in 2006. I am amazed at how much, often conflicting, advice is out there on the benefits or otherwise of feeding your cat a particular diet. There are websites recommending everything from a raw meat diet to a commercially manufactured vegetarian diet. Before you get lost in the maze of information and misinformation I will attempt to present your feeding options to you in a way that will not leave you dazed and confused. The important thing to remember is that many diets on the market are very good and that your vet is a great source of advice should you feel overwhelmed by all the choice.

What do cats need to eat?

Research over many years confirms that domestic cats thrive on a diet similar to that of their wild cousins of a comparable size. Don't be fooled into thinking that just because we can breed them with flat faces or curly hair we can turn them into chip-loving mini humans; we can't alter an animal's physiology. If you look at what and how a wild cat eats it will give you the basis of how we should feed our pets. Cats consume small prey whole. Larger prey is eaten over a longer period and certain parts, such as skin, fur, feather, larger bones and digestive tract, are left. Food is always consumed raw, which is why it is so important to ensure that any nutrients destroyed in the heating process of commercially produced food are replaced. We can break down the constituents of wild food and see what the cat's basic nutritional needs are to maintain health. A total of forty-one essential nutrients obtained from the following ingredients are required.

Protein

Cats, like other animals, need protein in their diet. Protein, once eaten, is broken down into specific amino acids (the building blocks of protein) that are necessary for the growth and repair of body tissue and regulation of the metabolism. As an obligate carnivore, a cat converts protein from a meat source into glucose (thereby creating energy), unlike our own body system, which mainly utilizes carbohydrate. It is unable to adjust this system of converting protein to energy, so this is one reason why a cat should not be fed a diet low in protein. There are certain specific amino acids that cats need in their diet – for example, cats are particularly susceptible to arginine deficiency (arginine is essential for metabolism). Taurine must be provided in a cat's food as it is essential for a number of body functions including the eyes and the heart. Nearly all taurine comes from meat, poultry or shellfish, so cats should not be fed a vegetarian diet. The recommended protein intake for an adult cat is at least 25 per cent of the daily calories consumed, although its natural diet is actually about 45 per cent protein, 45 per cent fat and 10 per cent carbohydrates and other nutritional elements.

Fat

Dietary fat is the most concentrated energy source of all nutrients; it also increases the palatability and texture of cat food. Fat also carries the fat-soluble vitamins A, D and E (see below) and supplies the essential fatty acids (EFAs) linoleic and arachidonic acids. These play key roles in maintaining the general health of the cat and are vital in many body systems, including the skin, kidneys and reproductive organs. It is recommended that at least 9 per cent of calories in the cat's diet should be provided by fat.

Carbohydrates

As mentioned before, the cat derives a great deal of its energy from protein, so it has no nutritional need for carbohydrate. It does, however, have the necessary enzymes to digest and metabolize carbohydrates, so they can form a useful dietary source of energy. Cats can therefore be fed wheatflakes and cooked rice, for example, although some cannot tolerate high concentrations of certain sugars. The value of carbohydrates in cat food is greatly debated among nutritionists but, despite this, most commercial dry foods contain between 30 per cent and 70 per cent carbohydrates. Essentially, by feeding these diets, we are meeting our cats' protein requirement with meat and their energy requirements with carbohydrates. This makes the food cheaper and enables the manufacturers to produce a cat food in dry form. Some health problems relating to the addition of carbohydrate in the diet include maldigestion (excessive gas, bloating or diarrhoea) and obesity.

Fibre is present as a source of carbohydrate in almost every commercial cat diet. It is referred to as an 'insoluble carbohydrate' because it is not digestible in the small intestine. The function of fibre in the diet is to increase both bulk and water in the intestinal contents, so it can be of benefit to sufferers of both diarrhoea and constipation.

Vitamins

Vitamins C, E and A are referred to as the antioxidant vitamins as they are important in preventing certain substances called free radicals from causing damage to cells and being involved in the ageing process. Vitamins may also be protective against certain forms of cancer.

Vitamin A Vitamin A is important for eyesight, the regulation of cell membranes and the growth of bones and teeth, but since cats are unable to synthesize vitamin A they must obtain it from organ meat such as liver and kidneys. Too much vitamin A can be as harmful as too little, causing hypervitaminosis A leading to lethargy, stiffness and skeletal problems. The daily requirement for an adult cat is in the region of 650–850 International Units (IU), equivalent to the amount present in only 5g of good-quality beef liver.

B-group vitamins These vitamins are essential for the conversion of food to energy in the body. Vitamin B1, or thiamine, is needed in relatively large amounts by the cat. This vitamin is added to commercial cat food because it is destroyed in the manufacturing process when heated. Even home-cooked meats would need to be supplemented with B1 for the same reason. Feeding your cat a raw fish diet may result in a deficiency of B1, owing to the presence of thiaminase, which destroys these vitamins.

Vitamin C Cats do not need to be fed vitamin C, as they are able to produce their own.

Vitamin D Vitamin D is involved in the metabolism of calcium. Animal tissue is low in calcium, so the cat's diet must be supplemented with this mineral. A deficiency of vitamin D results in rickets, although this is rarely seen in cats as they need very little vitamin D.

Vitamin E Although very uncommon, vitamin-E deficiency can occur in cats, particularly when they are fed food containing large amounts of unsaturated fats to which antioxidants have not been added. Unsaturated fats oxidize and go rancid easily and, as a result, the vitamin E present is destroyed. Yellow fat disease or steatitis is caused by a deficiency of vitamin E and may occur in a cat fed on red tuna, which does not have the necessary antioxidant or extra vitamin E added. Normal diets and proprietary foods containing tuna fish are adequately protected in this respect.

Minerals

Minerals can be divided into two groups: the major or macro minerals required in larger quantities (calcium, chlorite, magnesium, phosphorus, potassium and sodium) and the micro or trace minerals, which are required in much smaller amounts (arsenic, chromium, cobalt, copper, fluorite, iodine, iron, manganese, molybdenum, nickel, selenium, silicon, vanadium and zinc). Almost all (about 99 per cent) of the cat's body calcium and most of the phosphorus (85 per cent) is contained in the skeleton and teeth. Soft tissues such as meats and offal are very low in calcium and if they are fed as the sole food source calcium deficiency will occur. Good proprietary diets have adequate supplies of the major and trace minerals.

Water

Cats naturally produce concentrated urine, so their water requirements are relatively low in comparison with ours. The fluid content of a rodent or a tin of cat food (between 60 and 85 per cent) is often sufficient for most cats, but it is essential to provide fresh drinking water at all times. Cats fed on a dry diet need to drink water daily to make up for the lack of moisture. To encourage water consumption it is always advisable to provide water containers away from the food bowl; otherwise, you will find your cat naturally seeks out separate sources of water and will steal from the glass at your bedside table or hog the tap when you are brushing your teeth.

One point about water: many cats object to the smell of chemicals added to tap water, so boiled, bottled spring or distilled water may be used as an alternative. This is also useful when using one of the new pet drinking fountains in a hard-water area, as the motors will block with a build-up of limescale.

Understanding food labels

The more conscientious among us may have lingered in the cat-food aisle in our local supermarket to peruse the ingredients labels of our little darling's favourite nosh. There is an awful lot of stuff added to cat food and it's probably worth taking some time to study what it is and why it's there in the first place.

The Pet Food Manufacturers Association (PFMA) is the principal trade body in the UK representing the pet-food industry. The Association's aims are to promote safe pet food of sound nutrition and palatability that represents good value for money. They also represent their members' views to UK and EU government departments regarding pet-food legislation. All legislation applicable to pet food originates in the EU and is subsequently implemented in the UK. There are more than fifty pieces of legislation regarding the manufacture and sale of pet food, covering:

- labelling
- the use of additives
- contaminants
- use of animal by-products
- BSE (Bovine Spongiform Encephalopathy)
- selection of ingredients

The PFMA states that it uses those parts of the carcass that are either surplus to human requirements or are not normally consumed by people in the UK – for example, blood, bones, heads, feet and heart. It will process only those animals that the British generally accept in the human food chain – for example, they will not use horse, whale or other sea mammals. Apart from the meat content of the food there will be oils, fats, vitamins, minerals and carbohydrate fillers to provide bulk. The amount and type of carbohydrate will vary depending on the quality of the food. White rice is highly digestible for cats, for example, but other grains must be processed. Many cheaper cat-food brands contain large amounts of carbohydrate fillers, particularly in dry food. Unfortunately this is an essential ingredient to shape the dry-food pellets, so the quality of the carbohydrate is paramount. Some cat foods, to please us humans with our concepts of what constitutes 'good nutrition', advertise vegetable ingredients (e.g., chicken with carrots), but these have poor nutritional value for cats and are best utilized as a bulking agent for pets on weight-reduction diets.

Each pet-food label must indicate the species for which the product is intended, feeding instructions and 'typical analysis' of contents by percentage – e.g., protein 35%. Each ingredient must be listed in descending order of weight, so it's really important to check that meat is the first product mentioned. If moisture is present at a level greater than 14 per cent, then this has to be given as a percentage of the content also. This is why it is so important to compare foods by looking at a dry-matter analysis, as this varies enormously depending on the amount of moisture present in the particular food. If a food is called 'Beef Food' it must contain a 70 per cent or higher content of beef. If it reads 'Beef Dinner/Platter', etc., the manufacturer is required to include only 10 per cent beef. If the label says 'Cat Food with Beef' (sounding more yummy by the minute), then a mere 4 per cent of beef content is included. God forbid if the label reads 'Beef Flavour', as in that case there is no legal requirement for any of the food to be beef.

There is also an increase in the number of products that look like 'the real thing'; fish or meat in chunks that owners believe must be good for their cats because it is, after all, meat. Unless these products have been seriously enhanced you will find them labelled 'complementary' foods rather than 'complete', so these are not to be fed as a constant diet as they do not contain all the necessary nutrients.

Here are some commonly used terms for ingredients:

Animal by-products

These include unprocessed fresh or frozen material from a slaughterhouse and processed material, such as blood meal and meat and bonemeal.

Antioxidants

These prevent the growth of bacteria and are used as preservatives. Natural antioxidants such as vitamins C and E are often added and they also have nutritional benefits.

Artificial colours

This may list various E numbers or even names of certain dyes. Artificial colouring is linked to hyperactivity in humans but no such data is available regarding the cat's response to this.

Ash

This is a legal term used that represents the mineral content of the food determined chemically when the product is burned. The manufacturer is legally obliged to include this information on the pet-food label.

Beet pulp

This is dried residue of sugar beets from the sugar-production industry used as a source of fibre to aid digestion and prevent the formation of hairballs.

Beet pulp sugar, corn syrup, molasses

Cats are unable to taste a sweet flavour but sugars are often added to cat food to make the food moist and appealing in texture. There is a general rise in the incidence of diabetes in cats and this may well be linked to the increasing use of sugars in cat food.

BHA, BHT, ethoxyquin and propyl gallate

These are chemical preservatives best avoided. Unfortunately, they are often labelled 'contains EC permitted preservatives' or 'contains EU permitted antioxidants'.

Cereal by-products

These include wheat, rice, oats, maize and some sorghums (type of grass cultivated for grain).

Cereals

These are unprocessed whole grains, such as brown rice and barley, that are easily digested by cats.

Colouring agents

These act as preservatives.

Emulsifiers

These too are preservatives.

Fish meal
This is dried processed whole fish and fish offal.

Meat
By 'meat' any of the flesh is meant, including fat, skin and sinew in reasonable amounts, of any animal or bird normally used for human consumption, including, for example, head muscle meat, heart, pancreas, tail meat and tongue.

Meat by-products
This is the offal (e.g., liver, kidney, tripe) and blood, bone, heads, feet and some whole carcasses of smaller animals, e.g., rabbit. Named by-products in the ingredients, such as 'chicken by-product', is an acceptable constituent of the diet but shouldn't be listed as the number one protein source as it's not as well digested.

Meat derivatives
This is rendered carcass material. Rendering refers to the processing of raw carcasses on an industrial scale to remove moisture and fat. Some labels will state 'meat and animal derivatives (4% chicken)', for example. This indicates that meats such as chicken, lamb, beef offal etc. are combined to form the total amount stated. The 4 per cent chicken statement is a legal labelling requirement to show the minimum percentage content of the specifically mentioned meat guaranteed to be present by the manufacturer.

Plasticizers
These control the pH of food or modify the pH of urine.

This is by no means a comprehensive list but it gives you some idea of the minefield of complicated and often conflicting information that is out there. Always ensure you buy a cat food that has a help-line number listed for you to contact the manufacturer direct with any concerns you may have about the contents of their products. Most of the larger companies have websites that will guide you through their ingredients to give you a better idea of what you are feeding. Your vet will also be able to discuss any specific concerns you may have.

Feeding regimes

The cat's digestive system is adapted to frequent small meals throughout a twenty-four-hour period. Two large meals a day is hardly ideal, particularly as the average stomach capacity is about 0.3 litres, but somehow many pet cats adapt and cope with this regime. It is, after all, usually the most convenient feeding plan for owners out at work or with a busy lifestyle. In case you are undecided about how best to feed your cat, here are your options.

Feeding once a day

This is rarely the choice of owners; in a survey I conducted in 1995 only 1 per cent of those owners completing the questionnaire fed their cats once a day. This can, however, be an adequate regime if you are feeding dry food, as a bowl containing the day's ration can be left down in the morning. It's satisfactory, though, only if your cat isn't greedy and will take no more than what he wants when he wants it throughout the day. Many cats would gulp down a huge quantity of food in one sitting, regurgitate most of it and then wonder when they were going to get their next meal. I also think it's a potentially incredibly boring method as interaction during feeding time is often extremely pleasant for owner and cat alike.

Positive	Negative
• Convenient for owner	• Boring for the cat • Lack of interaction during mealtimes • Wet food would spoil, preventing attempts to feed later in the day

Feeding twice a day

This is the regime of choice for owners who feed wet food and are out during the day. The food spoils relatively quickly, especially in the warmer months, attracting flies and generally looking and smelling far too unpleasant to eat. If the cat isn't accustomed, or able, to eat all the food in one sitting, then a great deal of food is wasted. Cats rarely go hungry in these instances, as they soon get into the routine of adjusting their eating habits to ensure that sufficient is consumed first thing in the morning to prevent them from being too hungry before supper time.

Positive	Negative
• Convenient for owner • Able to feed wet food satisfactorily • Allows for positive interaction at mealtimes • Ensures cat's return at mealtimes • Predictable routine for the cat	• Can encourage competitive behaviour at mealtimes in multi-cat households • Wet food can spoil in hot weather • Still boring for indoor cats • Insufficient frequency for kittens and young cats • Need for strict timing as some cats find late meals stressful

Feeding three times a day

Kittens and young cats need to be fed at least three times a day. If you are at home during the day or working part-time, then feeding breakfast, lunch and dinner is always preferable to a twice-daily regime and gives your cat something to look forward to during the day. This is an ideal system for feeding an exclusively wet diet. If you are out at work during the day and would like to provide three meals, a dinner when you get home and a late supper as you go to bed may be the compromise.

Positive	Negative
• Ideal for kittens and growing cats • Enables wet food to be fed little and often • Prevents waste through spoilt food left down • More opportunities to have positive interaction with your cat	• Difficult to stick to routine times for meals • Strict mealtimes can be a cause of tension in multi-cat households

Constant food supply

This is the concept of 'ad lib' feeding to enable cats to dip into the food bowl throughout the day and night whenever they are hungry. The owner provides a bowl of food (only dry products are really suitable) and the cat chooses how much to eat and when. This best represents a natural feeding regime but is fraught with dangers if bored house cats turn to food for entertainment. It also encourages overeating, as owners rarely measure out the quantities fed and modern highly digestible dry foods are extremely palatable. If your cat has a tendency to overeat then ensure the amount put down daily does not exceed his requirements. If he consumes the entire amount in five minutes, it would be better to divide the amount into a number of portions and feed them at regular intervals.

Positive	Negative
• Enables a natural 'little and often' eating pattern • Lack of competition at mealtimes in multi-cat households	• Cats can overeat if the quantity is not monitored • No opportunities for food-related interaction between owner and cat • Wet food cannot be fed in this way

Feeding on demand

This was the feeding regime of choice for half of the owners surveyed in 1995. This relies on the cat 'asking' for food, hanging round the fridge or generally making it known that it is time for yet another quick snack. This sounds great in theory as it encourages intense owner/cat relationships. However, the cat becomes master and knows that he always gets what he wants when he wants it. Unfortunately the phrase 'give 'em an inch and they'll take a mile' could have been coined with cats in mind, as they are rarely satisfied with demanding food three or four times a day or tolerating it if you are asleep at 4 a.m. and unwilling to microwave that fish right now. There has to be a compromise here; if your cat is the laid-back type that is grateful for this regime and doesn't take advantage, then it's probably a good one. Beware, however, if you decide to adopt this regime from the start. There are potential pitfalls.

Positive	Negative
• Creates great owner/cat bonds • Prevents wasted food	• Cats can take advantage! • Could lead to excessive weight gain • Cat may not distinguish between daytime and night-time demands • Could result in a fussy eater as he discovers all good things come out of the fridge and not from a tin

Constant dry-food supply and twice-daily wet food

This, I believe, is the perfect compromise. The cat gets the variety of textures with both wet and dry food, the desirability of routine-structured mealtimes and the perception of abundance of food that a constant dry-biscuit supply provides. If the idea of being home at a specific time every night is depressing, then the modern timed feeders can be a good compromise. These enable a set amount of wet food to be accessed at a particular time with no risk of spoiling due to the ingenious use of a chill pack. This doesn't always guarantee that the meal will be at the desirable room temperature that most cats enjoy but one could argue 'What price structure and routine?'

Positive	Negative
• Creates owner/cat bond • Prevents wasted food • Alleviates the risk of tension at mealtimes as food is always available	• Can be difficult to regulate quantities • Lack of flexibility with the set mealtimes

How much to feed?

As well as the feeding regime options, there is still more that has to be considered. It's essential to get right the quantity fed, as weight gain in cats is potentially dangerous and as hard for them to diet away as it is for us. There are several factors to bear in mind, but all manufacturers are legally obliged to add feeding instructions to their food. These are great guides but they are not written in stone and certain factors need to be taken into account before gaily dishing out vast quantities of food that your cat doesn't need or want. At best you will waste a lot of money; at worst you will compromise your cat's health.

Here are some important factors to take into consideration when feeding your cat the optimum quantity.

- Your cat's lifestyle will influence its requirement for calories; feed less than the quantity recommended by the manufacturer if your cat lives indoors or is sedentary and sleeps most of the time.
- Kittens, growing cats, pregnant queens and lactating mothers with kittens need more calories.
- If dry and wet food is given in combination, then adjust the quantities accordingly.
- If feeding a multi-cat household, ensure each cat is obtaining its optimum nutrition; otherwise you may end up with one very fat cat and a few hungry ones.
- High-quality premium cat food is extremely palatable and cats may eat more than they need.
- Monitor any weight gain or loss once you decide how much to feed, as each cat's metabolism will vary; adjust quantities given accordingly.
- Titbits and treats represent a portion of your cat's daily

calories, so take this into account when calculating food quantities.

Food foraging

If you are feeding dry food exclusively or as part of the diet, then it can be very beneficial for your cat's mental and physical health to make obtaining these biscuits a little more challenging than 'removing them from a heap in a bowl in the kitchen'. A large part of a cat's natural endeavours to feed itself consists of foraging, exploring and problem-solving to catch the prey in the first place. Dry food can be secreted in various locations throughout your home, both on high and ground level. When you first start this regime your cat will probably sit and stare at you as you have always provided 'ready meals' in the past. Some cats are less capable of dealing with change in a predictable and routine existence, so this dramatic turn of events could prove unappealing. Once your cat has grown used to obtaining food in novel locations, the acquisition can become more challenging – for example:

- Place biscuits inside small cardboard boxes with the lids slightly open to encourage the cat to knock the box over or remove the food with its paw.
- Place biscuits inside commercially manufactured feeding balls and puzzles (available from most good pet shops).
- Place biscuits inside cardboard egg boxes with the lid partially shut.
- Place biscuits inside paper bags with the top folded shut.
- Place a couple of biscuits inside a screwed-up piece of tissue paper and throw it for your cat. He will pounce on the 'prey' and pull it apart to reveal the prize.

- Build pyramids of cardboard tubes from toilet rolls or kitchen rolls. Place five tubes side by side and glue together. Add four tubes on top and stick them together in a row and also stick to the tubes beneath. Continue to build with decreasing numbers in each row (5, 4, 3, 2, 1) to form a three-dimensional triangle. Place single pellets halfway along each tube and allow the cat to obtain the food by using its paw. Variety can be introduced by leaving some tubes empty and placing five or six biscuits in another. The bottom row of tubes is best stuck to a carpet tile to prevent the more lateral-thinking cat from just tipping the whole structure over and hoovering the biscuits up from the floor.
- Stick two yoghurt pots together to form a diamond shape. Make holes in the pots, approximately the size of a two-pence coin, using a soldering iron (this will ensure the edges are not sharp). Attach a piece of elastic through a hole in the top and hang about two or three feet off the ground. Place dry food inside and encourage the cat to tap and agitate the pots to obtain the food as it falls through the holes. To guarantee your cat's safety, make sure that the elastic has a breakaway section just in case he gets caught up in it during a frenzied attempt to get at the biscuits.
- Make holes in a clear plastic drink bottle (by melting with a soldering iron as above) and place biscuits inside for your cat to extract by rolling or manipulating the bottle.
- Some cats will enjoy chasing biscuits if you throw them across the floor. This is a nice opportunity to have some interaction during a feeding session.

Wet versus dry food

So much research is undertaken into flavour, odour and

texture of commercially produced cat food. Is a particular shape or size of kibble more attractive to cats? How important is the texture? Ironically, most of the research conducted on the odour of cat food shows that what the cat actually finds appealing is almost irrelevant. If it doesn't smell like a nice beef stew dinner to you, then it won't be marketed. Ironically, what the cat really likes may never hit the supermarket shelves, as it probably smells ghastly to us.

A number of things influence a cat's perception of 'delicious', including odour, taste, texture and temperature. A cat's sense of smell is extremely sensitive in comparison to ours and the odour of food plays a significant role in the initiation of feeding. If the food smells right, a cat may eat an otherwise bland diet. By far the most important factor to the cat is both the smell and the taste. Cats can taste food that's bitter, salty and sour, but they don't respond to sweet things; in comparison with us they have relatively few taste buds (cat=475, man=9,000). They are, however, sensitive and attracted to certain amino acids present in meat protein. The texture of food is also important, as cats tear their food before swallowing rather than chewing (watch your cat next time he attacks a large piece of chicken). The temperature too plays a part; cats like food best when it is at blood temperature (~35° Centigrade), as then all the subtle nuances of the flavour are released. This can be an important tip when encouraging reluctant cats to eat; heat the food a little and see if that helps.

Whatever you do to food and whatever variety you offer, the cat's taste will be influenced by both genetic and acquired feeding traits. It will vary according to the types of food it has experienced and any adverse response to it in the past. Cats also seem clever at rejecting foods that are nutritionally deficient in certain vitamins or minerals – for example, lacking in thiamine. It is often hard to get a cat to eat food that has a texture completely different from anything it has experienced in the past – for example, when introducing a wet food to a cat previously fed exclusively on dry food. This may be an important change to make, as some cats with recurrent cystitis are prescribed wet diets by their vets. The best way to introduce totally novel textural experiences is to mix the product gradually in increasing amounts with the original food.

So many other things can affect a cat's eating patterns, it's a wonder that some find the circumstances right to eat at all. The location of the feeding station may be too busy, near noisy washing machines or in full view of the tom from next door. The food container may be unpleasant, like, for example, some plastic bowls, or too near the litter tray. Physiological factors such as age and health can affect appetite, as can the stress of boarding, hospitalization, moving house and cohabiting with other cats.

Once we have satisfied ourselves that our cats are well and happy, we can, taking all the above into consideration, decide on what form we should present the diet in.

The pros and cons of wet food

There is a great deal of debate about the benefits of canned food versus dry food, but it is generally considered that the premium wet varieties are better because of their lower carbohydrate and higher water and protein content. Unfortunately, wet food is also more expensive, less convenient, smellier and messier. During warm weather, wet food spoils quickly and attracts flies, so it limits the number of ways that you can feed it to your cat without a lot of waste and a house full of bluebottles. Any cats with kidney or urinary tract disease certainly benefit from a wet-food diet, as they need higher water content in their food.

Positive	Negative
• Limited carbohydrate in comparison to dry • Provides adequate water • Predictable mealtimes usually required • Easier to eat for cats with no teeth • Nearer to texture and moisture content of natural prey • Easy to mix with any medication • Strong smell may encourage ill cats with poor sense of smell to eat • Bulk reduces risk of over-eating and aids bowel function	• Spoils quickly • More expensive • Smelly food/smelly faeces • Lack of flexibility with feeding regime • Allows tartar build-up on teeth

The pros and cons of dry food

Premium-quality dry food is palatable and easy to feed. It enables owners to be flexible with the pattern of feeding to entertain their pets. The kibble structure helps to reduce the build-up of tartar on a cat's teeth and usually results in fewer visits to the vet to have dental treatment if it is fed exclusively. However, it really is important to ensure your cat drinks enough water to compensate for the lack of moisture in the food. Dry foods are often very calorie dense, so, in combination with the high carbohydrate levels, obesity is a real danger. Dry diets have been implicated in the past in the development of urinary tract disease. Modern dry foods have been formulated to aid in the prevention of such problems, but the key to bladder health is ensuring your cat drinks enough water.

Positive	Negative
• Low cost • Convenience • Does not spoil so can be left all day • Flexibility of feeding regime • No smell • Preventative dental care • Reduces bulk in stool production	• Low-bulk can lead to constipation • Need to ensure the cat drinks enough water • Hard to mix medication with kibble • Risk of obesity due to dense calories and carbohydrate levels • Can lead to regurgitation if large volume swallowed quickly

As you can see, there is a great deal to consider. If you research the subject fully you will notice a lot of conflicting advice on the Internet. Many authoritative voices speak with passion and conviction but have completely opposing views. I would therefore suggest that, if your cat has no physiological need for a particular diet, you might hedge your bets on this one and provide him with 'the ultimate compromise'.

Feeding: the ultimate compromise

- dry food, in problem-solving receptacles, available throughout the day for challenging feeding opportunities
- wet food given as a tasty small meal twice a day

- feeding a wide variety of products to avoid 'addiction' or boredom
- providing water containers and pet fountains throughout the home, away from food bowls, to encourage drinking
- feeding the correct daily amount to guarantee optimum weight and health
- giving the occasional small treat (e.g., once a week) but taking its calorific value into account

Home-cooked diets

Many people are concerned about the nutritional merits of processed foods, both for themselves and their pets. The more proactive among them will investigate the possibility of creating a home-cooked diet. If this is something that you wish to explore, then it is important first to remind yourself of what a cat's natural feeding habits are.

The cat's diet is almost exclusively carnivorous, primarily consisting of small rodents, with some birds, reptiles and insects, depending on availability and seasonal variations. Rats and mice are often caught but not eaten; voles and young rabbits are the most palatable. Shrews seem to be the least palatable (possibly because they are insect-eaters), as they are rarely eaten. Therefore, whatever is fed within a home-made diet must provide all the constituents – including the appropriate balance of vitamins and minerals – of this natural diet. The added complication is that, being obligate carnivores, cats have a narrow range of tolerance for a number of dietary components, so deficiencies and toxicity are common if this balance is not achieved.

There are various arguments given by those individuals promoting home-cooked diets, one being that specific food hypersensitivities can only be established by initially feeding a home-made diet. Since some manufacturers have introduced 'novel protein source' and hydrolysed protein foods, this argument is no longer valid. The better-quality commercial pet foods are complete and balanced for all life stages of your cat from kitten to old age. Specially formulated diets are available to treat specific illnesses, combat the negative impact of indoor lifestyles and tackle the growing problem of obesity. All of these diets are produced under strict guidelines for quality and safety and are proved by extensive feeding trials. If you are still convinced that you could do better, there is a great deal of information in books on the subject or on the Internet to advise you. You will find recipes for Chicken and Catnip Jelly, Rabbit Risotto and even (not for the faint-hearted) Tuna and Marmite Milkshake. Unfortunately, not all sources of information on home-cooked food are good and some diets may be nutritionally unbalanced. Speak to your veterinary surgeon first and discuss your proposed recipes before getting down to the actual cooking. Here are just a few points to remember when considering home-cooked food for your cat:

- Table scraps are unlikely to cause nutritional imbalance if they represent less than 10 per cent of your cat's normal diet. They can, however, lead to obesity if you do not take into consideration their calorific value when measuring out the rest of your cat's daily allowance of food.
- A diet consisting exclusively of lean meat is not balanced as it has excessive amounts of phosphorus and deficiencies in sodium, copper, iron, iodine and vitamins.
- Raw fish can lead to a vitamin B1 deficiency if fed regularly.
- Cooked fish, if fed excessively, can cause a vitamin E deficiency, especially if it is packed in oil.

- Too much liver can lead to vitamin A toxicity, causing painful bone deformities.
- Onions (and garlic to a lesser extent) can cause blood problems, including anaemia.
- Grapes and raisins can cause kidney problems.
- Milk is not tolerated well by many cats as they lack the enzyme to digest it properly, causing diarrhoea.
- Green tomatoes can cause serious stomach upsets and even heart problems.
- Too high a proportion of fat can cause pancreatic problems.

Organic food

There are a number of products on the market that are advertised as 'organic'. This means that they are free from genetically modified foodstuffs, artificial colours and preservatives and are prepared using natural ingredients. A total of 95 per cent of the ingredients must be certified organic (the remaining 5 per cent must be from a permitted list) and flavourings must either be naturally or organically produced. Organically reared animals are not routinely given antibiotics or growth-promoting hormones. Crops are non-GM and grown without using conventional pesticides or artificial fertilizers. Those that support the use of organic pet food suggest that many digestive, skin and behavioural problems in pets could be due to chemical residues in their food. This has not, to my knowledge, been proved. Providing these diets are complete and contain all the necessary nutrients, then organic diets would be perfectly acceptable food for your cat. You will find, though, that they tend to be more expensive.

Raw meat diets

There is a growing trend in the UK towards feeding a raw meat diet. The first important point to make here is that an all-meat diet is not balanced and can result in serious deficiencies or excesses of certain vitamins and minerals. The emphasis in feeding 'raw meat' is that it should be meaty bones or carcasses that are fed as the bulk of the diet with some table scraps, cooked and raw, and constant provision of fresh water.

Appropriate foods recommended for this method of feeding include:

- chicken and turkey carcasses
- poultry by-products, including heads, necks, wings and feet
- whole fish and fish head
- whole carcasses, including rats, mice, rabbits, fish, chickens, hens, quail
- offal, including lungs, liver, heart, trachea and tripe (stomach)

The raw food should always be fed fresh. In warm weather it may safely be possible to feed your cat only once, in the cooler evening. The diet exponents recommend that table scraps be fed, either cooked or raw, as a proportion of no more than a quarter of the total diet and liquidized to aid digestion.

Things to avoid include:

- cooked bones
- excessive muscle meat
- excessive vegetables
- small pieces of bone
- excessive starchy food
- mineral and vitamin additives

- onions or chocolate
- processed food
- fruit stones
- milk

The whole diet regime is full of cautionary notes – for example:

- Many cats resist a change to a raw meat diet.
- Cats with dental disease or misalignment of the jaw may find it difficult to consume a raw meat diet.
- A wide variety of raw meat must be fed to avoid harmful excesses or deficiencies.
- Offal – e.g., stomach – must not exceed 50 per cent of the diet.
- Specific species of whole fish should not be fed constantly.
- Some species of fish should be avoided completely – e.g., carp can cause a vitamin B1 deficiency.
- Never feed pork; there's a risk of parasite infection.
- Avoid game meats for the same reason.
- To destroy parasites, keep the cat's food frozen for at least twenty days (although this makes it impossible to feed 'fresh' as recommended).
- Practise good hygiene to avoid parasite infestation.
- Exposing your cat to certain bacteria present in raw meat may be an issue.

If you have the stomach to prepare such meals, and your cat accepts them, then this is a real commitment in terms both of your time and your home. Feeding bowls, apparently, are unnecessary, so the recommendation is to feed the cat outside or on an easily cleaned floor, as the food will be dragged all over the place while your cat consumes it.

The most readily accepted alternative to a truly natural diet, by cats and owners alike, is a raw-meat-based diet of muscle and organ meat (meant for human consumption) supplemented and balanced with a variety of ingredients that can be purchased in a ready-mixed form. The manufacturer of this formulation does recommend, however, supplementation with whole prey, as the diet is soft and doesn't involve any shearing or chewing to consume. If your cat cannot be tempted to go out and kill something, there is a further recommendation of day-old chicks or chicken necks to tempt your cat.

Incidentally, I have treated a number of cats for wool- or fabric-eating, a behavioural problem with a genetic component linked to Siamese, Burmese and other derivative breeds. Many breeders of these pedigrees, when faced with this behaviour in their progeny, recommend that the owners feed day-old chicks and carcass meat. In my experience the cats just won't eat it or even chew it and the owners are highly resistant to feeding such a diet as, mostly, these cats are kept as indoor pets.

There are some very strong and sensible arguments for the theory of feeding raw meat diets. I totally relate to the concerns expressed at the high amount of carbohydrate that forms part of most commercially produced diets, and none of us is delighted about the use of chemical additives, colourings or preservatives, however safe they are purported to be in small quantities. My main reservation is in the execution. There are a lot of things to remember to ensure these diets are balanced. Most cats would need a long and tortuous weaning process from commercial food to a raw diet; some may never accept the change. I also believe that pet-food manufacturers have done wonders over the last ten or fifteen years by producing a number of diets that relieve symptoms and prolong good quality of life for cats and dogs suffering from specific diseases (although I'm sure that raw meat advocates would argue that the animal's diet may have contributed towards the disease in the first place). This would be extremely hard to reproduce in a raw meat diet.

There is ultimately no scientific evidence, only anecdotal, to support claims that raw meat diets are preferable to products that have been rigorously tested to confirm they are complete and balanced. My own feeling tends towards this being a problem of modern lifestyle and not the cynical commercialism of pet-food manufacturers. If cats lived in appropriate environments in appropriate numbers and were able to go outdoors and hunt freely we wouldn't be having this debate, would we?

Complementary foods

Complementary foods are those that, when fed alone, do not meet the nutritional requirements of the cat. Many of the 'gourmet' products of 'fish flakes' or 'meat chunks' do not actually satisfy all your cat's dietary needs. All complementary foods are labelled as such, with appropriate feeding guidelines, and should pose no risk for your cat if you feed them occasionally alongside a balanced and complete food.

Some supplements are occasionally recommended, but it doesn't necessarily follow that a substance has the same value for cats as it does for humans. The following foods should definitely be avoided.

Alfalfa powder, alfalfa sprouts Alfalfa, also known as lucerne (legume), is recommended in complementary human health as a tonic, antifungal, laxative, diuretic, detoxifier, with various other health-giving properties. However, veterinary literature lists it as a toxic plant for cats. Alfalfa contains cyanide-producing compounds that can be destroyed only by adequate cooking. Cats are not equipped to digest raw plant matter of any kind, no matter how thoroughly it is liquidized, and it will lead to severe indigestion. Cats certainly eat grass, but this is to trigger a mechanical process of regurgitation of indigestible or hard-to-digest matter, such as hairballs.

Yeast All the nutrients that make yeast a 'nutritional supplement' are readily available in muscle and organ meat as part of a normal feline diet. Yeast given as a supplement to a cat will start to ferment in its stomach, causing bloating and discomfort.

Vitamin C (ascorbic acid) Cats synthesize sufficient vitamin C (ascorbic acid) from glucose in their small intestine. Additional supplementation can be harmful.

Milk Cow's milk has little nutritional value for cats and will lead to weight gain if it is given regularly. Many cats are lactose intolerant and drinking cow's milk will cause diarrhoea or other digestive problems. Commercially manufactured 'cat milk' that is lactose-free is available and this can be fed occasionally as a treat, although it should still be noted that there is a calorific content.

Vitamins and minerals There is no need for vitamin or mineral supplementation if your cat is healthy and on a complete and balanced diet. Supplementation can lead to harmful imbalances and should therefore be given only on advice from a veterinary surgeon to treat a specific clinical deficiency.

Table scraps and titbits Like most things, table scraps and titbits are OK in moderation (for adult cats only), but please check pages162–4 for details of some human food that is highly toxic for cats. I would suggest that table scraps be confined to small pieces of meat, but don't offer them to your cat during mealtimes directly off your plate as this will soon lead to begging and, if your cat is bold enough, even 'stealing' from unsuspecting houseguests during dinner parties. If you regularly feed ham or cheese from your fridge you will soon find it impossible to get the door open without an urgent miaowing at your side. Commercially produced cat treats can be given as titbits but an allowance should be made in your cat's daily food quota; the same applies to popular goodies such as ham, cheese, tuna, chicken, yoghurt, butter, milk and cream. Strong-tasting or fatty substances such as Marmite, anchovy paste, paté, butter, cheese or sardines in tomato sauce can be extremely useful for disguising pills, but, again, the cautionary note is to use as small a quantity as possible as they are potentially high in fat, salt and preservatives and all have calorific values. Despite what you might think it isn't compulsory to give your cat something yummy every day; a treat should be just that: a rare but tempting morsel.

LIFESTAGE FOODS

Pet-food manufacturers understand that cats have specific nutritional needs during the various stages of their lives. Kittens, for example, for growth and energy requirements need roughly twice the nutrients required by adult cats.

Feeding kittens

Due to the small size of their stomachs, young kittens require frequent small meals (between four and six) spread at regular intervals throughout the day. By the time the kittens are twelve to fourteen weeks old, these meals can be increased in size and served in equal portions three times a day. At six months the kittens can be fed twice a day or allowed to 'graze' on dry food available throughout the day. This method of feeding will only suit those young cats that don't bolt whatever food is on offer.

Kittens should be weaned gradually from five weeks of age on to a variety of food, both wet and dry, specifically formulated for growth; this will be high in energy, protein and various key nutrients. A kitten's diet will also be easily digested and chewed. Exposing kittens to a number of different flavours and textures at this time will go a long way to ensuring they do not become fussy as adults or hooked on one particular variety. By the time a kitten is six months old it will have achieved 75 per cent of its adult body weight. Some manufacturers also offer an 'adolescent' diet that has a balance of nutrients to aid the development of bone and muscle that takes place during this period of growth spurt.

Feeding the adult cat

At the age of one year most cats (excluding the larger pedigrees) are considered to be fully grown and adult formulation diets are available to maintain body weight and condition. At this point, and thereafter until your cat becomes elderly, you can also choose a diet that is specifically tailored for his lifestyle or particular vulnerabilities – for example:

- indoor cat (for less active cats with a tendency to form hairballs and gain weight)
- outdoor cat (for cats with 'extra' energy requirements)
- hairball control (for cats with a tendency to form hairballs)

- light formula/weight control (has lower fat/calories to help prevent weight gain)
- multi-cat (with a little of everything to appeal to a variety of different tastes)
- sensitive skin (for those cats prone to itchy skin)
- sensitive stomach (for those cats prone to stomach upsets)
- dental care (with larger kibble to scrape plaque off teeth)
- hypoallergenic (free from wheat, soya, dairy, beef)
- urinary support (helps to produce acidic urine)

One manufacturer even offers formulae for specific pedigrees – for example, Persians and Siamese – maintaining that their unique attributes require a unique diet. Once again, I would suggest that you should consult with your vet before chopping and changing to include every fad on the market.

Feeding the older cat

You will find more on this in Chapter 9, but essentially the older cat also has specific dietary requirements to maintain a healthy condition. Diets tend to boost the immune system and contain highly digestible fat and protein. Many of the larger pet-food manufacturers stage diets to match the ageing process, including one that is formulated for the 'active mature' cat (7+ years) and one for the 'senior' cat (11+ years). Senior foods are recommended for cats from seven years and upwards, depending on the company.

Feeding the pregnant or lactating queen

After mating, queens (female cats) need more calories, peaking between five and six weeks after the kittens are born. A growth formula is recommended for female cats at this time, but your vet will give you specific advice regarding feeding protocols.

Lifestage products vary enormously; some have such small quantities of the 'special' added nutrients that they could have no real impact on the animal's health. If you want to ensure you are feeding the right diet to your cat at the right time in his life, it's worth investing in those products that have proven efficacy in proper feeding trials. If you are still in doubt, ask your vet; he or she may say that your cat is well maintained on its existing diet and any changes to expensive alternatives with extra bits added is unnecessary.

Hypoallergenic is one of those words that are often used misleadingly. In the wording of non-prescription diets that claim to be 'hypoallergenic' you will find that they omit certain foods, such as beef, that have been known to be linked with allergies and contain a protein source that is relatively uncommon in commercial cat food. However, most 'hypoallergenic' foods contain lamb, turkey, rabbit and duck and these are readily available in most cat-food products nowadays. Your vet is best equipped to decide if your cat needs a hypoallergenic diet and, if necessary, he or she will recommend an appropriate prescription diet that addresses the problem.

Prescription diets

Prescription diets are a relatively new innovation in the world of pet nutrition. One single manufacturer is the leading producer of diets, sold exclusively through veterinary practices, that are designed to treat and alleviate the symptoms of a number of chronic conditions and diseases. The list currently includes diets for:

- colitis

- constipation
- convalescence
- dental disease
- diabetes mellitus
- gastrointestinal problems associated with dietary sensitivity
- heart disease
- hyperlipidaemia
- kidney disease
- liver disease (including hepatic encephalopathy and copper storage disease)
- lower urinary tract disease
- obesity
- pancreatitis
- skin problems associated with dietary sensitivity
- weight loss associated with metabolic disease

Finally, a few words on prescription diets for urinary tract disease. The most common presentation of this disease in the UK is referred to as feline idiopathic cystitis. It is a stress-related illness with complicated causes and a wide variety of symptoms. Studies show that the most effective dietary change that an owner can make to manage this condition is to feed a wet food to increase the cat's fluid intake and thereby reduce the concentration of the urine.

Vegetarian/vegan diets

I'm afraid I have to include this section in the chapter on food for your cat because there is so much information on the Internet on this very subject. There are some convincingly written arguments for providing your cat with a non-meat diet that is heavily supplemented with the nutrients, such as taurine, that are found naturally in the appropriate quantities only in meat.

There are at least three companies that manufacture and market vegetarian and vegan food for cats. One actually proudly proclaims 'not tested on animals', which worries me somewhat. Needless to say, they are not members of the PFMA (Pet Food Manufacturers Association). On searching the Internet I was dismayed to see that the well-known and outspoken charity PETA (People for the Ethical Treatment of Animals) is accusing people of jeopardizing their animals' health by feeding them commercial pet foods and it is actively encouraging a change to a vegan diet for their cats.

I would suggest to those people offended by the sight or smell of meat products that they choose a proprietary dry cat food. I cannot emphasize enough that cat vegetarianism is absolutely the wrong dietary choice. We are all entitled to hold strong views regarding the consumption of meat in human society, but it is irresponsible in the extreme to impose these beliefs on our cats. Even the Vegetarian Society itself is cautioning against it. We already restrict our cats' lifestyles enough; let's not add this lunacy to the list.

Bad foods

There are few foods that are inherently poisonous to cats, but a certain number of very common human foods will, if too much is fed, cause illness and even death. Here is a checklist of the most common (some of which appear in the section on home-cooked diets):

- all-meat diet – causes nutritional secondary hyperparathyroidism
- chocolate – contains theobromine causing vomiting/diarrhoea, muscle spasms, seizures

- cod (fish) liver oil – if fed excessively will cause hypervitaminosis A and vitamin-D toxicity
- cooked fish – if fed excessively will cause vitamin-E deficiency
- grapes, raisins – toxicity mainly reported in dogs but best avoided
- green sprouted potatoes – contain solanine causing gastrointestinal disturbances
- green tomatoes – also contain solanine causing acute gastrointestinal disturbances
- liver – if fed in excessive quantities will cause hypervitaminosis A
- mushrooms – some reports of poisoning
- onions, garlic, shallots, chives – Heinz body anaemia
- raw egg – may cause dermatitis, hair loss and neurological dysfunction
- raw fish – vitamin B1 deficiency
- tuna – if fed exclusively will cause a form of osteoporosis

Other potentially dangerous foods include:

- alcohol
- apples and apricots (stems, seeds and leaves)
- avocados
- baking powder and baking soda
- cherries
- citrus-oil extracts
- coffee, tea and other drinks containing caffeine
- dog food in large amounts
- elderberry
- fatty foods
- macadamia nuts

- milk and other dairy products (many cats are lactose intolerant)
- nutmeg
- peaches and plums (stems, seeds and leaves)
- rhubarb (leaves)
- salt in large quantities
- sugary foods
- tobacco
- yeast dough

CHAPTER 5

Keep Young and Beautiful!

GENERAL MAINTENANCE AT HOME

ONE QUALITY THE CAT POSSESSES THAT I GREATLY ADMIRE IS its ability to look good at just about every time of the day or night. Its extraordinary flexibility enables it to groom places we couldn't even dream of getting to without the aid of a long loofah. However, in order to keep our cats in tiptop condition there are times when a little help is necessary. They may get poorly and require medication or have dental disease and need their teeth cleaned; there will always be occasions in the life of pampered domestic cats when humans can give a little helping hand.

You could consider this to be the chore-based side to cat ownership, but, remember, many situations are man-made, so we really are responsible. For example, we breed cats with coats so long and lustrous (and unnatural) that the poor mites couldn't stop it knotting unless they kept completely still their entire lives. We confine cats indoors but don't want our little predators to sharpen their claws on our furniture, so we trim them short. We give them gourmet food instead of a dozen mice to chew, so they get bad teeth. Do you see my point?

On a positive note, if our cats are brought up to accept the sort of handling that general maintenance of their health and good looks requires, this affords quality time when owner and cat can do a little bonding. Unfortunately, it's not so 'hearts and flowers' for some. Let's hope the following advice keeps the stress to a minimum.

Grooming

Most cats will spend a significant part of their day grooming. Their bodies are incredibly supple and can bend and flex to enable them to reach all parts with ease (which is not always the case when they get older – see Chapter 9). The cat's tongue is covered with backward-pointing spines, perfectly designed to groom coats effectively, removing loose hair and dirt.

A normal cat's fur is made up of three types of hair: down, awn and guard. Most of the hair (approximately three-quarters) is down and the rest are the more protective awn hairs, both growing in clusters from single follicles. The few guard hairs that are present grow singly. Some pedigrees' coats are made up of different proportions of these three types of hair, giving them their distinctive coat types. Most house cats moult to some extent all year round and grooming helps them to remove the dead hair, much of which is swallowed. This hair becomes impacted and either passes through your cat's system or is vomited up as a hairy brown sausage on your dining-room carpet. This isn't one of the cat's most endearing habits.

Grooming performs several important functions. It:

- removes loose hair and smooths the coat to help insulate the body more efficiently
- regulates temperature in hot weather by spreading saliva across the coat that subsequently evaporates, cooling the cat down
- keeps the coat waterproof by stimulating glands at the base of the hairs
- spreads sebum along the coat, producing, when exposed to sunlight, vitamin D, which is subsequently ingested by the cat
- spreads the cat's own scent across its body (often cats wash after being touched or stroked to enable them to taste whatever they've come into contact with and to re-establish their own scent)
- removes parasites
- maintains strong social bonds when cats groom each other

Benefits of grooming your cat

Grooming your cat, irrespective of its coat type, has a number of benefits. It improves muscle tone and stimulates the skin to produce oils, giving the coat a healthy shine. Your cat will remove a large amount of hair himself when grooming, but if you get involved on a daily basis, particularly when he is moulting, you can remove a fair amount of the hair and prevent the formation of hairballs in his stomach. Grooming also gives you a chance to give your cat a quick physical, looking at ears, eyes and mouth and checking for fleas, ticks, ear mites or any lumps or bumps. The whole process can be very therapeutic for both of you and is a positive way of reinforcing the bond between you.

Start as you mean to go on

You should get your kitten accustomed to being groomed from a very early age. Gradually build up the amount of time you spend grooming your kitten until it is quite happy to allow you to brush it. Eventually the kitten will enjoy being groomed all over and will come to see it as part of its daily routine. If you approach grooming as a chore, then so will your cat. It may even put up a bit of resistance, resulting in grooming becoming a task you put off. If your cat gets used to being handled in kittenhood it makes it easier for the vet to carry out an examination. Your cat will know what to expect and there will be no nasty surprises. It also makes any visits to the surgery less stressful for both you and your cat.

How to groom shorthaired cats

Equipment: fine-toothed flea/tick comb, bristle brush, rubber grooming pad or mitten

- Groom at least once a week.
- Use the flea comb first to check for flea excreta (tiny black comma-shaped dirt that runs red if a drop of water is added to it). If you find any, see Chapter 3 for details on flea control.
- Use your fingers to loosen the dead hairs in the coat by massaging against the hair growth, in the direction of his head.
- Gently brush or comb your cat's hair, using strokes in the direction that the hair grows.
- Use the bristle brush to sweep the coat in the direction of the head, and then smooth it down again (some cats don't like this bit if they are not used to being brushed in this way).

- Use the rubber grooming glove or pad from head to tail, removing dead hair and stimulating the blood supply to the skin.
- Loose hair lying on the surface of the coat can be removed with a damp cotton or rubber glove.

How to groom longhaired cats
Equipment: wide-toothed comb, rubber mitten or pad

- Longhaired cats need to be groomed at least once a day. Don't even consider acquiring a Persian, for instance, unless you have the time to devote to this task for the rest of his life.
- Before grooming, massage the skin thoroughly by rubbing with your fingers against the hair growth from tail to head. As the hair is lifted, check for flea excreta (see above).
- Using a wide-toothed comb, groom from head to tail to remove dead hair.
- Take particular care with areas under your cat's 'armpits' and between his hind legs, as the skin is very thin here and extremely sensitive. It is also an area of friction where he is likely to have knots.
- When you encounter a mat or knot, tease it apart using your fingers, working from the root towards the end of the hair.
- Avoid the use of scissors; it is extremely difficult to see where the skin ends and the hair starts when it has become really matted and you will inevitably cut him at some point.
- Check the hair between the toes and pads for mats. Any accumulated debris can be teased out gently.
- Use the rubber mitten or pad to remove more dead hair.

- Remove dead hair on the surface of the coat with a damp cotton or rubber glove (or your hand).
- Finish off with the comb again.
- If grooming is a struggle, try offering food treats and talking reassuringly. Begin grooming when your cat's attention turns to the treat.

There are numerous grooming products on the market, particularly for longhaired cats – for example, rakes, slickers and detangle sprays, many of which claim to make grooming as simple and safe as possible. What works for you and your cat can be a very personal thing, so it's worth investing in a range of equipment until you find the combination that does the job and causes the least amount of stress for all concerned.

You can also purchase one of a number of freestanding or wall-mounted grooming aids that encourage your cat to rub against them, thereby removing dead hair. It is not an alternative to grooming but it certainly can be pleasant for your cat.

Dematting Unfortunately, sometimes it just isn't possible to keep longhaired cats completely knot-free and they develop mats that can be extremely uncomfortable (put a clump of your hair in a tight elastic band that constantly pulls at the scalp and you will experience what this feels like). You may find a professional cat groomer who would still be able to groom manually but most severe mats need to be shaved off by your vet. This will require at the very least sedation; most likely it will be done under general anaesthetic. You will then find that your cat ends up with fur on his tail, head and lower legs and nowhere else until it all grows back.

Bathing your cat

If your cat is healthy there is no reason to give your cat a bath. Most do not tolerate it well and routine bathing is probably best reserved for the show cat.

There are occasions, however, when it may be necessary; my own cat Mangus, for example, has a skin condition that requires regular medicated baths. This actually takes place at the veterinary practice as they have the necessary equipment. Mangus is far less likely to struggle there than at home (after all, where would she go if she escaped?) and I would rather she didn't have negative associations with the place in which she feels most secure. Fortunately, and unlike most other cats, Mangus enjoys the attention of her baths; the water temperature is always just right, all towels are warmed beforehand and even the medicated shampoo is heated to body temperature. Not your average bath at all, it's more like a visit to a beauty salon.

If you do have to bathe your cat (maybe he's had an unfortunate toilet accident), you will probably find it a struggle, as most cats detest the business of being restrained and covered in water or soap. You will probably need to wash only his back end, or at least only part of him, rather than immerse him totally. It's not a good idea to bathe him regularly as this strips the skin of the natural oils that waterproof and insulate the coat.

If it's absolutely necessary, here are a few tips:

- Prepare your bathroom beforehand to make sure everything is within easy reach.
- Put the heating on in there and shut the door once you have your cat *in situ*.
- Use the bathroom basin or bath.

- If you have a shower attachment for the taps this will be extremely useful for rinsing.
- Put a rubber non-slip mat in the bottom of the bath/basin.
- Groom your cat first (if possible) to remove any knots in the area you intend to bathe.
- Wet the area thoroughly with water that is set at body temperature.
- Use only cat shampoo, not something for your own hair!
- Massage the shampoo in well.
- Rinse thoroughly until the water runs away clean and all the shampoo is out of his coat.
- Gently squeeze out excess water and remove your cat to a warm towel for drying.
- Have extra towels available to continue drying and replace the damp ones.
- Some cats will tolerate a hairdryer, but keep it on a low setting and move it constantly to prevent it from overheating one area of skin.
- Keep him in a warm place until he is completely dry; he will probably groom himself frantically at this point.
- Once the coat is dry give it a final groom.

If you manage this single-handedly, *congratulations*! Probably, though, you will need a willing assistant to give you both encouragement to get through the ordeal. Don't worry, your cat will learn to trust you again some day . . .

Brushing your cat's teeth

The best way to avoid dental disease becoming a major problem for your cat is to indulge in a little daily preventative care.

Plaque and tartar can build up on your cat's teeth and this may lead to infections developing at the gum margin. Gums eventually recede over time and the teeth loosen. Dental cleaning and extractions are performed under general anaesthetic at the surgery; this involves ultrasonic descaling of the teeth, the removal of any that are damaged or loose and a final polish. It is an expensive process that can be avoided if, once again, you start as you mean to go on and get your kitten used to having his teeth brushed regularly. Some dry diets claim to reduce the build-up of plaque and tartar on your cat's teeth, but this is not a complete substitute for slightly more hands-on dental care. Dental chews and gels are also useful; your vet can advise you on these. So, no more excuses, let's get brushing!

When you first acquire your kitten, as well as enjoying all the fun stuff you will be handling him to get him acclimatized to the sort of fiddling he will encounter at the vet's surgery or during any other general health-maintenance activity. Teeth brushing is one of those activities, so your kitten should learn to accept gentle teeth and gum rubbing.

- Always begin the process when your kitten is relaxed and quiet.
- Start by rubbing your index or middle finger gently round your kitten's mouth.
- Once this is accepted readily, gently insert your finger into the mouth by slipping it in, pointing towards the back of the head, at the crease where the upper and lower lips join.
- Move the finger slightly backwards and forwards, up and down, along the margin of the gum to get your kitten used to the movements associated with brushing.
- If this is tolerated, try to use a toothbrush specially

designed for a kitten's mouth. Do not use human toothbrushes or toothpaste!

- Once the kitten is used to the toothbrush, you can introduce toothpaste that is designed for cats and usually is flavoured to taste meaty or fishy. Pick a flavour that your kitten enjoys.
- As your kitten grows up, make sure the toothbrush you are using is the right size.
- If your kitten or cat never gets used to a toothbrush it is better to brush using your finger and cat toothpaste than not to brush at all.

If you are attempting to brush your cat's teeth for the first time when he is already adult you will not find it quite so straightforward. Make sure you choose the appropriate time; it's no good starting when your cat has inflamed gums, as they will be sore and this will put him off brushing for life. It's probably best to begin as a preventative measure for the future after your vet has treated any dental conditions and your cat is feeling more comfortable. With an adult cat you will get the best results if you give him a treat afterwards and keep the sessions short to start with. You may only manage to use your finger and toothpaste, but be sure always to keep your finger parallel to the teeth to avoid being bitten accidentally.

Administering liquids and syrups

Occasionally your vet may prescribe important medications that come in liquid or syrup form. This may be more acceptable to those cats that struggle with tablets, but the flavour is often geared to the paediatric patient and cats don't particularly like, for example, 'strawberry'-flavoured anything! You may also have occasion to give food in liquid form to your cat, so it's important to know how to give it properly without risking his inhaling it by mistake.

Here is a step-by-step guide to make the process as stress-free as possible:

- Tell your vet straight away if you are unable to give medication to your cat; there may be alternatives.
- Ensure the product is stored appropriately – e.g., in the fridge.
- Check the dosage before administering and shake the liquid if necessary.
- If your cat is really resistant to medicating, check to see if the liquid can be mixed into his food. If so, warm the food slightly first to release the flavours and, with any luck, disguise the taste of the medicine.
- If it can't be given in food, then place your cat on a table top with a non-slip surface.
- Talk gently to him and try to relax yourself.
- If you have assistance, one person can gently hold his front legs and the other can administer the liquid.
- If you are on your own, wrap his body in a towel.
- If he isn't a wriggler or a struggler, the least amount of restraint the better.
- Support your cat's head in a slightly raised position and place the tip of the syringe or pipette in the cat's mouth, behind the canine tooth (the long fang) and pointing towards the opposite side, not down the throat.
- Press the plunger on the syringe (or squeeze the bulb of the pipette) slowly to allow the liquid to trickle on to his tongue.
- Your cat may chew at the syringe; don't worry, though, because that will help him swallow.

- If he coughs at all, stop the liquid and lower his head.
- Once he has received the full dose, if your vet says it's OK to do so, give him a little treat.
- Keep your cat indoors to guarantee continuity of medication and always complete the course.

Giving tablets and capsules

Most of us have seen the joke instructions for 'giving your cat a pill'. They involve all manner of appalling injuries to the owner and very little drug intake for the cat, and just about every cat owner can relate to it. When I worked as a veterinary nurse I spent a lot of time pill-popping into unsuspecting cats because they were absolutely impossible to dose at home. I developed a technique used by most veterinary nurses that was so quick and non-invasive that the cats didn't have time to object. Unless you are administering pills every few hours to your cat you are unlikely to develop this skill, so here is some advice that may help you:

- It is much easier to give pills to a cat that is used to being handled round the head and mouth. If you are acquiring a kitten, ensure it has become accustomed to this.
- First, check to see if tablets can be crushed, or capsules opened and powder sprinkled, into food.
- Some capsule contents lose their bitter flavour if they are kept in the fridge, so consult your vet about this.
- If you are putting powder or crushed tablets in food, then warm the food (make sure it's something that is highly palatable) or mix the medicine in soft butter first and put the butter in the food. Other suggestions include mixing the tablet or powder in ham, soft cheese, Marmite,

anchovy paste, sardines in tomato sauce or tuna.

- If the tablet or capsule has to go down the throat unaided, try to do it as gently and quickly as possible; 'less is more' when it comes to restraint.
- If you have an assistant, one person should hold the front legs and the other give the pill. If you are on your own, use a towel to wrap the cat in.
- If you are single-handed, tuck his body under your left arm (if you are right-handed), facing forward, and support him gently using your left elbow.
- Place the thumb and index or middle finger of your left hand (if right-handed) either side of your cat's head on the corners of his mouth.
- Tilt his head upwards until his nose is pointing to the sky; the lower jaw will become slack and open slightly.
- Hold the tablet between the thumb and forefinger of the right hand and use the middle finger to gently open the jaw by pulling down on the incisors (the small row of teeth at the front).
- Drop the tablet to fall on the back of his tongue.
- Close his mouth, continue to hold the head up and gently stroke his throat.
- If you are struggling to get it down, coat the tablet in butter or margarine and try again.
- Some tablets can dissolve in the oesophagus, so give your cat some watery food after he's swallowed the pill (check with your vet first).
- Keep your cat indoors to guarantee continuity of medication and always complete the course.

Administering eye drops

You may need to give eye drops to your cat to treat one of a number of conditions. They are extremely effective and it's really essential that your cat receives the right amount at the necessary interval.

- Store the drops according to the manufacturer's instructions.
- Shake the drops if recommended to do so.
- Ensure the appropriate eye is treated.
- Wipe any discharge from around the eye with damp cotton wool.
- Place your fingers on top of your cat's head and your thumb under his chin to support the head.
- Tilt the head back slightly and open the eye, if necessary, by raising the skin on the head (and therefore the eyelid) by pulling back your fingers.
- Drop the required number of drops on the surface of the eye and allow your cat to blink before lowering his head.
- Ensure the tip of the dropper does not touch the eye. It is safest, if you can, to keep the heel of your hand in contact with the head while you manipulate the dropper: this ensures that you won't accidentally poke him in the eye if he jerks his head suddenly.

Administering eye ointment

Proceed as above, but squeeze the tube, holding it away from the eye, and allow the ointment to be deposited on the surface of the eye or on the inside of the lower eyelid. Close the eyes and massage the skin round the eye with thumb and forefinger to spread the ointment.

Administering ear drops or ointment

Ears are very sensitive things, so don't be tempted to stick a cotton bud down your cat's ear or generally fiddle with anything beyond where you can easily see. Always leave the delicate stuff to the expert: your vet! You may be asked to administer ear drops or ointment at some stage, so this is the simplest way:

- Shake the bottle and store the product according to the manufacturer's instructions.
- Clean any discharge from the outer surface of the affected ear with damp cotton wool.
- Hold your cat's ear tip and slightly elevate it with one hand, then administer the required number of drops into the ear by holding the bottle upside down with the other hand and squeezing it.
- Massage the base of the ear by squeezing the sides together several times to allow the drops to enter the ear canal.
- Wipe away excess drops from the outside of the ear.

Trimming your cat's claws

Trimming your cat's claws will protect your furniture and prevent any accidental scratches on your skin but should be considered only for those cats kept exclusively indoors. Once again, starting as you mean to go on will always make life easier for you. Getting your kitten used to having his paws gently fondled and massaged will reap benefits later in life.

There are several types of nail clippers that are suitable for trimming your cat's claws. The claw should be inserted into the clippers laterally (the flat side inserted parallel to the clipper blades) and cut gently to avoid crushing the nail. Make sure any clippers used have sharp blades. Here we go:

- Always reward your cat with a tasty treat afterwards and choose a time to do it when your cat is sleepy rather than just after a stressful experience or boisterous game.
- Have your cat resting comfortably on your lap, the floor or on a table.
- Hold one front paw in your left hand (if you are right-handed) and press the toe pads gently to extend the claws.
- You should be able to see (when you view the claw side on) the pink tissue inside the claw that extends from the base towards the tip of the claw. This is the quick of the nail that contains the blood and nerve supply. You want to avoid cutting near to this as, if you damage it, it will be extremely painful for your cat and it will bleed. I would recommend you trim about halfway between the end of the quick and the tip of the claw.
- You may find your cat a little restless to start with (it may even be because you are a little uptight), so don't worry if you only trim one or two claws at each sitting. Try again later when you are both more relaxed.
- The finished result should last for up to four weeks before more trimming is necessary.
- Cats rarely do damage with their hind claws as they maintain them by chewing. If they are particularly long or your cat scratches himself with his hind claws and causes harm, you may want to ask your vet or a professional groomer to trim them.

Plastic nail caps

If your cat scratches itself or your furniture and you can't trim the claws yourself for whatever reason, you may consider soft plastic caps that can be fitted over your cat's trimmed front claws by your vet (hind claws if self-harming). These will last six to eight weeks before they need replacing and your cat will not harm himself or your furniture during that time. Again, though, this is not recommended at all for cats with access outdoors.

Alternative and complementary therapies

There is a growing interest in alternative therapies and treatments in human medicine. It is not uncommon nowadays for your GP to refer you to a 'complementary medicine practitioner'. As we embrace the whole concept of homoeopathy, acupuncture and osteopathy we also want to know that our pets can benefit from the same wide choice of treatment protocol should they get sick.

The Royal College of Veterinary Surgeons has now recognized the effectiveness of manipulative therapies, such as physiotherapy, in the treatment of animals. In the veterinary profession's Guide to Professional Conduct it states that vets must show 'accountability, accessibility and transparency' when informing their clients of all the treatment options. It's also important to point out that we have a duty of care to our animals, so we shouldn't really consider alternative therapies for our cats. You or I can decide what weird and wonderful things we do to our bodies in the name of curing illness or promoting wellness, but we must seek the best possible treatment for our animals to prevent suffering. Therefore, whatever we decide to do I strongly believe that we should do it with the full knowledge and blessing of our vets. If we want to give flower essences or nutritional supplements we should consult them.

There really is no alternative to veterinary treatment but there may be therapies that complement it.

In the United States in 2006, a survey was conducted to investigate owners' use of complementary medicine for pets suffering from cancer. Of those surveyed (254 clients), 65 per cent used complementary and alternative therapies such as herbs, nutritional supplements, chiropractic care and acupuncture. Sadly, many owners did not tell their vets that they were giving this treatment or seek their advice about it. It is worth noting that, apart from chiropractic, osteopathy and physiotherapy, all therapies, including acupuncture, homoeopathy and aromatherapy, must be undertaken by veterinary surgeons. Even the manipulative therapies can only take place if a vet has diagnosed the condition and decided it's appropriate. Admittedly, by law, you (the owner) can give whatever treatment you want to your animal providing it doesn't involve intrusion into its body, e.g., an injection. This is, of course, subject to the Protection of Animals Act 1911 and the subsequent Animal Welfare Act 2006, which oblige you not to cause suffering. Properly trained and qualified animal manipulative therapists are members of their respective professional bodies and are covered by indemnity insurance that protects you should anything go wrong.

There is a misconception that all herbal and homoeopathic remedies are 'safe'. Unfortunately they can interfere with the action of conventional treatments and drugs and animals can have reactions to them. There is no legal requirement for them to be tested in the same rigorous way as pharmaceutical products. I have complete empathy with owners with extremely sick pets and I understand the need to 'try everything'. When my cat Annie was suffering towards the end of her life with crippling osteoarthritis I consulted a veterinary chiropractor to provide her with some relief from her pain. I did this, though, with the full knowledge and blessing of my veterinary surgeon. Unfortunately the faith in complementary medicine is often to the exclusion of the conventional type and I believe this is where we can store up problems for ourselves.

A quick A–Z of complementary therapy

Acupressure

Acupressure is based on the same principles of traditional Chinese medicine as acupuncture (see below), but involves no needles and can be done by you, the owner, providing you familiarize yourself fully with the technique.

Traditional Chinese medicine considers that a flow of energy (chi or qi) sustains and nourishes the body and if this flow becomes blocked in any way it can lead to illness. Chi circulates along invisible pathways throughout the body called meridians. There are specific points along those pathways where, if pressure is applied, blockages can be cleared to restore harmony.

Acupressure is applied by the hands and fingers; the choice of acupressure point depends on the condition from which your cat suffers or the specific aspect of his health you wish to address.

Acupressure is recommended for:

- pain relief
- healing of injuries to increase blood supply and remove toxins
- strengthening muscles, tendons and bones
- joint lubrication
- reduction of swelling and inflammation
- enhancing the immune system

- restoring physical and emotional energy

If this is something that appeals to you I would recommend you contact a veterinary acupuncturist who may be able to show you how to perform these techniques at home. Certainly the calm nature of acupressure and the gentle touching would do no harm and may well enhance the bond between you and your cat.

Acupuncture

Working on the same principle as the acupressure, acupuncture uses needles to stimulate the specific points along the meridian. Western medicine has embraced this treatment in recent years and there are very different approaches between the two cultures to interpreting its mode of action.

The Chinese approach is holistic, taking emotional, genetic and environmental factors into consideration when treating disease. Traditional Chinese medicine aims to achieve balance physically, emotionally and spiritually to restore health.

The Western scientific approach has come up with three theories that may explain how it works.

1 Polarized electric-like energy fields exist within and around the body and these can be influenced by external electromagnetic fields. The acupuncture points are like amplifiers and the meridians are the conductors of the bioelectricity (the qi). If the regulation of these bioelectric energy fields becomes disturbed, the insertion of the acupuncture needles in the relevant points will enable the system to self-regulate itself back to normal. Clear as mud?
2 The second theory suggests that acupuncture releases specific chemicals – for example, endorphins (a potent morphine-like substance) – into the blood, which can alleviate pain. On this basis, acupuncture can stimulate the

immune system and treat harmful inflammatory reactions in the same way.

3 The neurophysiological hypothesis is that the acupuncture points are located in areas where there are concentrations of peripheral nerves. Accurately placed needles could therefore communicate directly with specific receptors that in turn send a message through the nervous system. Using this theory it would be necessary to know what to 'turn off and on' in the animal's nervous system to address the specific problem – for example, stimulate bone healing or inhibit pain.

Acupuncture for the treatment of animals in this country is largely used to alleviate musculoskeletal disorders and chronic pain. Conditions in cats that respond to treatment include:

- back pain – e.g., spondylitis, spondylosis and disc disease
- arthritis
- strains and sprains
- chronic diarrhoea or constipation
- neuritis causing itching and overgrooming
- urinary and faecal incontinence
- epilepsy
- feline asthma and other chronic respiratory conditions
- chronic kidney failure
- dysfunction of the immune system
- stress-related disorders – e.g., aggression (in conjunction with behaviour therapy)

Acupuncture is not just about fine sterile needles. There are occasions when heat is also applied (moxibustion) or low-voltage electricity (electroacupuncture) and sound (sonopuncture). Tiny sterile gold or silver beads can be implanted on acupuncture points or even a small quantity of vitamin B12 can be injected. Recent innovations in the application of acupuncture also include auricular therapy, where the acupuncture points are focused on the animal's ears. These points can be used for both diagnosis and treatment.

If you feel that acupuncture may be helpful in treating your cat, then the first person to discuss this with is your vet. He or she will make the call and decide whether the treatment is appropriate. If you are referred to a veterinary acupuncturist you will probably have to take your cat for a course of treatment, but signs of improvement are often seen after a couple of sessions. Cats receiving acupuncture treatment will often relax totally once the needles are in place and even go to sleep; they can be quite drowsy afterwards.

Acupuncture is considered to be a successful form of treatment for many conditions, but it has its limitations. It doesn't cure cancer and it won't reverse a degenerative disease process once it has gone so far. It is useful for relief of symptoms in certain cases, but it won't, at this stage, fix your cat.

Aromatherapy

Essential oils are potent extracts of aromatic plants, the chemical constituents of which are easily absorbed via the nose or through the skin. In human aromatherapy, massage is the primary method of application. Most people have quite a cavalier attitude to aromatherapy: just a few nice smells that make you feel good. However, according to those in the know it is an effective and powerful form of medicine that can cause serious harm if used incorrectly. Some oils can cause abortion in pregnant animals and excessive use of tea tree is suspected to cause hormonal disruption in young boys.

Veterinary surgeons who practise alternative veterinary medicine will often use aromatherapy oils in the waiting and consulting rooms to calm the patients. A cat can be treated directly with an aroma using a bowl of steaming water, with the oil added, placed in front of the cat carrier with a towel draped over both.

Common aromatherapy remedies include basil to aid the digestion, eucalyptus as an expectorant, lavender for its calming effect and tea tree as a disinfectant. They should be used with caution, though, as they can clash with homoeopathic remedies and conventional medicines used to treat the same condition.

The Veterinary Surgeons Act 1966 restricts aromatherapy treatment of animals (other than your own) by anyone other than a fully qualified vet. This doesn't mean I would recommend that any owners take it upon themselves to experiment. Cats are extremely sensitive and should never have essential oils applied topically as they have a limited ability to metabolize them.

Flower remedies

Flower remedies are made from extracts of numerous plants and flowers and they are used to treat undesirable emotional states and physical illnesses caused by stress. The principles behind their mode of action are similar to those of homoeopathy, working through vibrational energy encouraging the body to heal itself. The treatments were originally devised for use in humans but they are said to be equally effective for animals, especially cats.

The best known of all the flower remedies is the range developed by Oxford-based homoeopathic physician Dr Edward Bach well over sixty years ago. Flower essences or remedies have been around for hundreds of years; they are used by many cultures, including the Egyptians, Africans and Australian Aborigines.

The remedies can be used in combination, up to a maximum of five, or individually. For cats they should always be diluted prior to use and administered in food, drinking water, straight into the mouth or even on the back of the neck or paw.

These remedies are readily available in health-food stores and chemists, but, once again, I would always recommend that the vet be consulted before use. Since each remedy is specifically created to treat a particular emotional state or illness, it requires an accurate diagnosis of that condition before treatment can begin. Many owners misinterpret their cat's mood state and might well, therefore, use the incorrect remedy.

I have, with the blessing of the vet concerned, quite often suggested that these remedies be used alongside behaviour therapy for various problems. There are a number of other factors that become relevant when these remedies are used, such as a reduction in owner stress because something is being done, increased contact between owner and cat during administration of the remedies and a general level of expectation of efficacy of the treatment. These, in my opinion, go a long way towards improving the situation for owner and cat alike. I have certainly heard a great deal of anecdotal evidence regarding flower essences and, even if I don't understand the science behind them, I am prepared to keep an open mind.

Herbal remedies

A number of vets throughout the UK use herbal remedies alongside conventional veterinary medicines. They can be either licensed veterinary herbal medicines or herbal supplements. Herbal medicines are licensed under the same rigorous controls as any other drug and efficacy must be proved in order to produce them to treat certain conditions. Herbal supplements are not governed in the same way and must be sold merely as additions to the animal's diet of the relevant vitamins, minerals or compounds.

Traditional medicines generally extract and concentrate one active chemical from a plant, whereas herbal medicines utilize the whole plant. Supporters of herbal medicines believe that, as the plant contains many other chemicals alongside the active one, the combination may concentrate its effect, make it easier to digest or absorb or lessen its harsh or toxic side effects.

There are two main types of herbal remedy: Indian and the more familiar European. The Indian herbs follow the traditions and principles of Ayurveda, a system of health care that predates traditional Chinese medicine. These herbs are given to balance the body, as illness is seen as a sign that the natural balance has been disturbed. The particular herbs are chosen based not only on the symptoms but also on the physical and psychological make-up of the individual being treated, referred to as the 'Dosha' type. This is hard enough to assess in humans (it takes years of training) but almost impossible in animals. Therefore vets tend to recommend the herb that treats the symptoms without reference to the Dosha type. Indian herbs can be purchased in syrup form for cats and a wide range is available to treat a variety of ailments.

Homoeopathy

Homoeopathy is a system of medicine that follows the principle of treating 'like with like' to cure illness. The medicines used can be made from any source – animal, vegetable or mineral – such as sand, arsenic, charcoal, salt or certain insects, and some more recent remedies are derived from man-made substances. Those substances, which would have the same effect on a healthy body as the symptoms seen in the sick human or animal, are then used in minute quantities to treat the patient.

An essential part of the process is attention to the diet to ensure the recipient of the homoeopathic remedy is eating as naturally and healthily as possible. Under the terms of the Veterinary Surgeons Act 1966, it is illegal for anyone other than a vet registered with the Royal College of Veterinary Surgeons to prescribe homoeopathy for animals or to diagnose or give advice based upon a diagnosis.

The homoeopathic consultation is incredibly detailed. In order to prescribe the appropriate remedies it is necessary to find out as much about the patient as possible – not just the condition now but signs and symptoms shown at any time in the past. The homoeopath needs to find out about the patient as an individual: its likes and dislikes, reaction to stimuli, its environment, diet and all other aspects of the animal's care are analysed thoroughly. I often feel that this methodology is why alternative and complementary therapies are so popular. Owners finally find a practitioner who has the time to listen! I consulted a veterinary homoeopath during the latter stages of Spooky's illness (Spooky was the first cat I owned). The process was very gentle, calm and reassuring and I really felt I was doing everything when conventional veterinary medicine had done all it could. She didn't get better but I do think there was a little relief from her symptoms towards the end.

Correctly and accurately prescribed, homoeopathy can be safely used on animals of any age and works alongside most conventional treatments without compromising either. It can be particularly useful before and after surgery to relieve the psychological stress and promote rapid healing and reduce discomfort.

Your veterinary surgeon will refer you to a veterinary homoeopath if he or she feels it would be appropriate and helpful for your cat's condition. I honestly don't believe it should be the primary source of medical treatment for cats, but it has an important place in providing a complete and thorough programme of care.

Physiotherapy

Physiotherapy refers to the use of physical techniques for the treatment of injuries and movement dysfunction. In human medicine it has been proved to be indispensable in the recovery of musculoskeletal conditions and it is now recognized within the veterinary profession as being equally useful in the rehabilitation of many injuries in horses, dogs and cats.

The most common injuries that respond to physiotherapy are:

- osteoarthritis
- back pain
- muscle atrophy (wastage)
- joint, tendon and ligament injuries
- paralysis
- hip dysplasia
- wound management

The physiotherapist will use a number of techniques to achieve a reduction of pain, improvement in mobility and increased flexibility. These include:

- soft tissue and joint mobilization
- acupressure (see above)
- ultrasound
- hydrotherapy
- Transcutaneous Electrical Nerve Stimulation (TENS)
- pulsed magnetic field therapy
- massage
- specific exercises

Some may think that hydrotherapy – i.e., swimming in a pool to aid mobility and speed recovery from injury or surgery – is not really a cat thing. This should be judged on a case-by-case basis and many hydrotherapy centres have seen great improvement in the condition of cats utilizing this form of physiotherapy. A veterinary nurse friend of mine recently visited such an establishment and she marvelled at the dedication and enthusiasm of the staff there. When she posed the question 'Do you have many cat patients?' she was told that they had two felines who regularly visited for treatment. After a shaky start the two cats were paddling around madly, but, to respect the sensitivity of the species, they were allocated separate ends of the pool, out of contact with each other. However, the rule soon became relaxed when it was apparent the cats wanted joint activity and the remainder of their therapy was conducted with both of them swimming full lengths of the pool and constantly looking across at the other as if competing in some Olympic race. Such, it would seem, is the competitive nature of cats!

Psychics

The one thing above all others that my work with animals has taught me is that it's important to keep an open mind and to explore the merits of every form of therapy available to the conscientious cat owner. That doesn't mean all types are valid or helpful, it just means that all should be examined for evidence of efficacy without the hindrance of preconceived ideas.

Animal psychics (or animal communicators) are individuals who claim to be able to 'tune in' to an animal's thoughts and feelings and translate them into a language that humans can understand. They offer assistance to owners to achieve a number of goals, including:

- the diagnosis and resolution of behavioural issues
- the relief of physical and emotional trauma, grief, guilt or fear
- contact with animals who have died, to express the love of the grieving owner
- tracking down lost or missing animals

I don't wish to challenge anyone's spiritual beliefs but I have as yet been unable to find any concrete evidence in support of this as a valid form of therapy. I am not suggesting for one moment that there is any cynical motivation for animal psychics to do what they do, but a great deal of their 'tuning in' amounts to simple cold reading techniques and vague comments that any owner would want to hear or would certainly find it difficult to prove or disprove.

Maybe there is something in it. I just don't know.

Reiki

Reiki comes from the Japanese words *rei*, meaning spirit, and *ki*, meaning energy, the translation being 'universal life energy'. It is a form of healing that is channelled through the hands of the Reiki practitioner. It can be used alongside conventional medicine and other therapies, as it is non-invasive with no side effects. It can be used to relax the healthy or reduce pain and anxiety in the sick. It also claims to improve the effectiveness of other treatments.

This is another very spiritual form of alternative therapy, but I wouldn't hesitate to recommend it to anyone who felt his or her cat would benefit. It will certainly do no harm and the gentle and calm nature of the treatment is relaxing for both the patient and the owner. Once again, I don't know how it works, but any potential relief of symptoms and anxiety in a sick or dying cat has got to be worth a try. However, if the practitioner works from a clinic rather than in your home it's important to weigh up the possible negative effects of a car journey against the positive outcome of the visit.

Supplements or nutraceuticals

'Nutraceuticals' are food supplements, in the same class as vitamins, but specifically designed to nutritionally support certain illnesses, improve metabolism, correct imbalances or enhance performance. They are not manufactured as pharmaceuticals and so are not subject to, or controlled by, the same legislation as drugs.

Aloe vera This thick-leaved plant is used for animals much as it is for humans, as a soothing, anti-itch ointment applied directly to the skin. It is also said to help heal cuts and protect them from infection.

Biotin One of the B group of vitamins, biotin is thought to enhance the body's ability to absorb nutrients. It promotes healthy skin, coat and nails and is sometimes given to cats suffering from itchy skin or allergies.

Brewer's yeast This is a source of protein and B vitamins to maintain normal body functions and stimulate appetite.

Cod-liver oil capsules These are recommended for joint mobility and to improve coat condition.

Echinacea Echinacea stimulates the immune system and helps animals fight off infections and diseases.

Elderberry and nettle extract This is a preparation that enhances coat condition.

Evening primrose oil This is often recommended in the treatment of various skin conditions to improve the health of the skin and coat.

Ginger Chinese medicine uses ginger as an anti-inflammatory and an aid for stomach problems. Ginger is said to help animals with car sickness and digestive problems such as flatulence and diarrhoea.

Ginkgo Humans take it to increase their memory and improve their brain function. Some vets are using it to treat animals that exhibit cognitive dysfunction, the animal equivalent of senile dementia.

Glucosamine Glucosamine is a naturally occurring substance produced in the body that is used in the maintenance and regeneration of healthy cartilage in joints. Glucosamine sulphate and chondroitin sulphate are effective treatments for degenerative joint disease such as osteoarthritis. It has also been recommended for the treatment of idiopathic cystitis in cats, on the principle that it aids the repair of the damaged bladder-wall lining. Recent research, however, has shown that it has little therapeutic effect on many sufferers of this condition.

Kelp seaweed powder This maintains a healthy metabolism and enhances coat and skin condition. It is often recommended for animals with pica (see p. 318) as it contains all the essential minerals and trace elements.

L-Carnitine This is recommended for the maintenance of normal cardiac muscle and skeletal muscle performance.

Milk thistle Animals with liver problems may be given this supplement to protect the cells of the liver from toxins.

Slippery elm This is used as an aid to the digestive system for pets that suffer from constipation and upset stomach. It has also been used as a cough suppressant.

St John's wort This herb is used in human medicine as a treatment for depression, but in alternative veterinary medicine it is used as a treatment for viral infections and nerve disorders.

Valerian This has a calming effect and is often combined with skullcap as an anti-anxiety supplement.

Wheatgerm oil It maintains normal fertility in breeding animals and promotes a healthy nervous system and skin and coat condition.

Some herbs can interact with prescription medicine, so check with your veterinary surgeon before administering any nutraceutical or supplement. Your vet is there to help you and your cat; he or she will probably be delighted to see you are so keen to do the very best for him.

Tellington TTouch

Linda Tellington-Jones created Tellington TTouch as a teaching method for animals, using non-familiar exercise and body movements to increase confidence, reduce tension and generally enhance the relationship between animal and owner. The practitioner will, for example, move the cat's skin using circular movements of the fingers or hand to release tension and increase mobility.

TTouch claims success in treating a number of fears and phobias – for example:

- dislike of physical contact
- lack of confidence
- car sickness
- noise sensitivity
- fear biting
- timidity

Any owner can train in TTouch or even just read a book about it and practise some of the basic techniques. It is becoming very popular in rescue centres to assist in the handling of difficult or anxious cats. If your cat is resistant to handling or nervous at the vet surgery, this could be something you could use to good effect.

It is impossible to keep up with all the new fads in the world of alternative and complementary therapies. If you come across anything that isn't mentioned in this chapter, then find out as much as you can about it with particular reference to research that has been conducted into its effect. Don't just believe that it is safe and it works because the person selling it tells you it does. Once you are equipped with all the information available, talk to your vet about it. There may be a time in your cat's life when conventional medicine can only go so far and you are searching for something to relieve pain and suffering. If you have your vet's blessing that a treatment will not harm your cat, then my advice would be to give it a go. It's a horrible feeling to think that you haven't done everything you possibly can.

CHAPTER 6

Home Comforts or the Great Outdoors?

LIFESTYLE CHOICES

THE LATEST STATISTICS SUGGEST THAT AT LEAST 10 PER CENT of the nation's cats are kept exclusively indoors. A further significant percentage has restricted access to the great outdoors when their owners are there to supervise them. I am always speaking to owners who feel they have no alternative but to restrict their cats' access outside owing to the massive amount of danger they perceive is there. It is certainly true that an alarming number of young cats are killed on the road. Anyone who has lost a loved pet in this way would understandably be reluctant to risk such a thing happening more than once.

Whether we like it or not, though, it is impossible to

guarantee our cats absolute safety and at some time we will have to make the monumental decision 'Home comforts or the great outdoors?' In an ideal world no cat would have restrictions on its movements and no cat would ever be bred that was so modified it had to be confined for its own safety. However, the reality is very different. Some cats don't want to go outside and some homes have no easy access, so there will always be circumstances in which the only solution is to keep the cat exclusively indoors.

Home comforts

It would be impossible to reproduce accurately a full and natural lifestyle indoors. This doesn't mean, however, that we shouldn't try as best we can to simulate certain elements of an outdoor environment so that our indoor cats have an opportunity to behave as nature intended.

Unfortunately the modern trend towards minimalistic home environments is not the best news for the nation's favourite pet. Where are the multiple levels? Where are the nooks and crannies? Where's the excitement? In an ideal cat-friendly room a cat could travel from one side to the other without ever touching the carpet: from armchair to sideboard to bookcase to occasional table to footstool to sofa . . . great fun!

Regrettably, if a cat becomes bored, frustrated, depressed or generally stressed in a sterile and unstimulating environment it is possible that he may indulge in behaviour that we consider to be highly undesirable – for example:

- aggression towards people or other cats
- urination or defecation in inappropriate places
- urine-spraying
- over-grooming or fur-plucking

- overeating leading to obesity
- anxiety and depression
- over-attachment to owner
- attention-seeking behaviour

If the problem is not addressed it can also lead to the development of disease. It is worth noting that all these potential problems could be prevented if we took a little time and trouble to adjust our homes to suit our cats' needs as well as our own.

The easiest way to ensure your home is cat-friendly is to look at all the provisions contained within it that are purely for your cat or used by your cat. These items or facilities provide nourishment, entertainment, stimulation and a sense of security. They are your cat's favoured 'resources' and the things that you need to ensure are present in the right quantity and quality – for example, feeding 'stations', litter trays, private areas and toys. If you have more than one cat, the quantity is particularly important. Ideally each resource would be supplied following the formula: one per cat plus one extra in different locations; for example, in a three-cat household there would need to be four separate areas where food is available.

Food

It is very tempting to leave food down all day in a bowl or feed our cats tinned food twice a day. If we are trying to mimic the cat's natural eating habits, then we are failing dismally! A 'natural' diet would require hunting for several hours a day and the consumption of as many as a dozen small rodents. The average pet cat potentially has a lot of spare time to fill with other activities, often sleep. It doesn't have to be this way as the possibilities are endless for exciting feeding opportunities if your cat is eating dry food (unfortunately it just doesn't work with wet food); see Chapter 4 for further details.

Water

If you are using the dry-diet food-foraging approach it is essential that there is every opportunity for your cat to drink. However, positioning water near food can actually deter some cats from drinking sufficient fluid and this could lead to urinary tract disease. Finding water elsewhere can be extremely rewarding for your cat and there should be at least two water containers positioned in different locations. Some cats object to the chemical smell of tap water, so filtered or boiled water can be used. There are various ways to provide water, including:

- *Pet water fountains* There are an increasing number of these available, some of which filter the water. They do need regular maintenance, though, to keep the motor from clogging in hard-water areas.
- *Feng shui water features* While these are not designed specifically for cats, they can be just as entertaining as the genuine pet fountains if you stick to the simple 'water running over pebbles' type.
- *Ceramic, glass or stainless-steel bowls* Keep these topped up to enable the cats to lap from the edge yet still see the surface of the water. Plastic bowls can smell quite offensive to cats, so these are best avoided.

Vegetation

A source of grass is essential for the house cat as it acts as a natural emetic to rid him of any nasty hairballs. It can be purchased as commercially available 'kitty grass', or pots of grass and herbs can be grown indoors specifically for this purpose. I discuss later in this chapter poisonous and cat-friendly house plants, so it's worth checking since some cats will chew other plants, even if grass is available, just out of curiosity.

Litter trays

The positioning and type of tray is crucial for the indoor cat as we are basically taking away his right to choose where he goes to relieve himself. Each cat will have his preferences but there are a few rules that will help you get it right:

- The locations should be discreet and away from busy thoroughfares.
- Trays should be located well away from feeding areas or water bowls.
- Ideally give your cat a choice between covered and open trays.
- Avoid positioning them adjacent to noisy household appliances – e.g., washing machine.
- The trays should be cleaned regularly.
- Some cats are averse to polythene liners and litter deodorants.
- The litter substrate should reflect the cat's natural desire to eliminate in a sand-like substance.
- Never expect an indoor cat to share a tray with another.

Social contact

Don't forget the role you have to play in entertaining your house cat; social contact is considered to be a highly desirable resource providing the quality and quantity is right. This may seem harsh but not all cats want you in their faces twenty-four hours a day.

You really should allow your cat to dictate when and how all this social stuff takes place. It's not just hugging and squeezing; it can be play, grooming, talking, even running around and chasing him or playing hide-and-seek (two games that seem to be very popular with many owners and cats alike).

Social contact is not just with humans, so when you first acquire a kitten and decide it will be kept as a house cat it would be kinder to consider getting his brother or sister too. There is always a slight risk that they will have problems as they get older, but the provision of the right number of resources within the home will guard against the need to compete. It is also important to remember that company can come in different forms, so don't forget the dog.

High resting places

Cats are natural climbers and it is important that the home environment reflects this by providing opportunities for them to rest and observe proceedings from an elevated vantage point. This will encourage essential exercise and is particularly important in a single-storey home without stairs. Any places provided should be located in such a position that the cat is able to get down; it is always easier to climb up. Here are some suggestions for suitable locations:

- Tall scratching posts are available as modular units and there are even floor-to-ceiling structures for the super-adventurous. These are best anchored to the wall in some way, as it is virtually impossible to make these rigid and free from the odd wobble.

- Free-standing cupboards and wardrobes are popular but try to avoid encouraging your cat up there, which may look suspicious and probably deter him from choosing this as a safe place.
- Shelves can be constructed specifically for your cat's use, or else merely clearing an area on an existing bookcase would do just as well. Wooden shelves can be slippery, so the provision of a non-slip surface would be ideal.
- If you really want to create an adventure playground, then secure a section of loop-pile carpet to a wall and this will represent a challenging climbing frame. It can be fixed by attaching double-sided adhesive carpet tape to a clean wall surface. The carpet is then stuck to the back of the tape and wooden batons are positioned at the top and bottom (secured with screws and rawlplugs) for added security. It may be advisable to have shelves or cupboards nearby to enable your cat to come back down without too much trouble.
- A heavy-duty cardboard tube from the inside of a roll of carpet can be utilized indoors. It makes an ideal climbing frame if it is secured and covered with a layer of sturdy carpet.
- If you want to appear rather stylish and avant-garde, acquire a selection of second-hand high-backed chairs and fix them to form a stepping ladder up the wall and nice places to rest up high. This sounds bizarre but actually looks quite striking!

Private areas

Secret places are a must in any cat household to give your cat an opportunity to rest in absolute security. Areas under the bed, in the airing cupboard, at the bottom of the wardrobe, and so on, all represent excellent hideaways. Do remember to respect your cat's desire for privacy and don't interrupt him while he's there.

Beds

You can go mad buying expensive beds but your duvet and the sofa will always win hands down. Anywhere in the sun or a radiator hammock are also popular. Cats prefer to sleep off the floor, so if you do succumb to the charms of a leopard-print bed make sure you place it on a high surface to give it maximum appeal. Electric heated pads are a great purchase if you want to encourage your cat to frequent a particular area. If you cover it with a thick fleece, the lure of the soft, warm surface will be irresistible.

Scratching posts

Cats need to scratch to maintain their claws and mark their territory, so if you don't provide scratching posts or panels your furniture will suffer. Some cats like horizontal surfaces and others prefer vertical; you may have to experiment to see what your cat prefers. Make sure whatever structure you choose is rigid to withstand a strong pull from a set of claws and tall enough to let your cat do it at full stretch.

Catnip

Two-thirds of cats respond to the smell of the herb catnip (*Nepeta cataria*). Cats rub their faces on it, roll in it, eat it, scratch on it and generally get highly excited or extremely mellow. It can be such a useful thing to promote play and even to encourage appetite if it's sprinkled on a novel food. Catnip toys do lose their potency after a while, so storing them in a self-sealed polythene bag containing a pinch of dry catnip will keep them going for a lot longer.

Predatory play

Fishing-rod toys are ideal to simulate the movement of prey. They can be agitated in front of your cat and he will eventually be unable to resist the desire to pounce. A huge variety of these toys is available commercially but you can make one of your own with a bamboo cane, strong adhesive tape, a little string or elastic and a feather or two. Always make sure these are not left out for excitable individuals to drag around the house, as they may get tied up in knots if left unsupervised.

Toys

Many cats enjoy retrieval games and these can represent an opportunity for social contact as well as play. Elasticated towelling hair bands are just the right size for a cat to pick up and kittens are particularly quick at learning the game of 'fetch'.

You don't have to be on the end of a toy to make it fun, so it's important to have a range of things that your cat can play with when you're not at home. I've already mentioned this in Chapter 3, but it's worth repeating here, as toys are an important part of life for the bored house cat. Anything made out of real fur or feather will always be a winner because it taps into a very primitive hunting drive.

Novelty is always the key to getting your cat excited, so

leaving his toys lying around all day is boring. Put one or two toys out daily, maybe placed inside something to give your cat the task of removing it with his paw, and keep the rest in a cupboard to bring out randomly.

As recycling is an important part of everyday life there is no harm in recycling rubbish into cat toys! The following items are suitable candidates:

- a rolled-up and scrunched piece of paper or tin foil
- a cork
- the plastic seal on the top of a milk container
- the plastic lid off a tube of crisps
- the plastic seal off the tub of pasta you had for lunch
- cardboard boxes
- paper bags
- supermarket carriers (cut off the handles to avoid your cat rushing about with a bag round his neck)
- a walnut (they make a great sound)
- an empty crisp packet tied into a knot
- sweet wrappers
- a piece of ribbon, shoelace or string

All these items are probably best reserved for playing when you are watching, but don't think that because they cost nothing they won't be greatly appreciated. My Mangus has a box of posh toys but she's a rubbish girl at heart, much preferring sweet wrappers and hair elastics.

Novel items

Even if you provide all the above in great abundance there will always be occasions when your cat finds everything predictable and boring. A degree of familiarity and predictability is good but your cat must be challenged and stimulated to really get the best out of life. From his point of view anything that comes through the door is worthy of investigation: the soles of your shoes, the weekly shopping and even the wheels of your baby's buggy. Smell is so important to your cat and he will love the opportunity to explore new things. Bring home bits of wood, stone, cardboard boxes, bags . . . anything that he can sniff and examine thoroughly. He will never shout *'I'm bored!'* but it's probably wise to presume he is to some degree, so keep trying all the time to think up new exciting things for him to see and do.

Fresh air

Don't forget fresh air is a great source of information and entertainment for your cat. This is absolutely essential if you are a smoker, as house cats can suffer physically and emotionally from a constantly poor atmosphere. There are a variety of grills and meshes that can be fitted to a window opening that allow air in but prevent the cat from falling out.

High-rise syndrome

While on the subject of fresh air . . . This isn't necessarily what you want to hear but it's fascinating nonetheless. There is a phenomenon in the cat world that occurs every time a cat falls from a balcony or window. The resulting trauma, in cases of falls of over two storeys, is called high-rise syndrome. Cats that fall from between two and thirty-two storeys have an overall survival rate of 90 per cent. Those that fall from a height under six storeys have more severe trauma than those falling greater distances. Typical injuries include:

- chest – rib fractures, pneumothorax (air in the cavity that surrounds the lungs)

- facial/oral – fractured jaw, fractured hard palate, broken teeth, tongue and head trauma
- limbs – fractured fore and hind legs
- abdomen – damage to the liver/spleen/kidneys, ruptured bladder

The theory why cats tend to fare better when falling from greater heights is interesting. It has been calculated that cats reach terminal velocity after about five storeys, at which point they relax, allowing the force of impact to be distributed over a wider area. If a cat lands before reaching maximum speed it tends to be rigid and the full force of the landing impacts on the part of the body that reaches the ground first. What amazing creatures!

Poisonous household items

Falling out of windows is not the only risk the house cat faces. There are plenty of things inside the home that are potentially dangerous, particularly those that, if ingested, could be poisonous.

It is probably true that cats are less likely to be poisoned than dogs. They tend to be fussier eaters, so, unlike some dogs, they won't tackle any old thing that's vaguely organic, but there are characteristics of the species that make them susceptible in other ways. Many cases of poisoning in cats occur as a result of a toxic substance getting on to their coat; cats will automatically groom it off, thereby ingesting the poison. They have a small body size, so just a tiny amount can cause toxicity; also, a cat's liver doesn't break down chemicals very well.

Cats can ingest prey that contains poison, absorb poison through the skin or even inhale it. It isn't incredibly common, so don't panic, but it's worth getting to know what constitutes 'poison' for your cat and what you should do to help swiftly. A cat has a much better chance of making a complete recovery if veterinary treatment takes place as rapidly as possible.

Signs of poisoning

The clinical signs are very variable, depending on the poison involved. Some toxins have an impact on more than one body system, so there may be a combination of the signs listed below:

- gastrointestinal signs – vomiting and diarrhoea
- respiratory signs – difficulty in breathing, coughing, sneezing
- neurological signs – tremors, lack of coordination, depression, excitability, seizures, coma
- dermatological signs – inflammation, swelling
- kidney failure – increased drinking, loss of appetite, weight loss
- liver failure – vomiting, jaundice (yellow mucous membrances)

This is the action to take:

- Remove your cat from the source of poison if appropriate, taking the necessary precautions to keep yourself safe.
- If the contamination is minor – a small amount of oil on your cat's fur, for example – and your cat hasn't groomed, then the most effective way to remove the risk is to cut the section of fur off. Placing a comb between the skin and the scissors will prevent any risk of cutting the skin.
- If the toxin has significantly contaminated your cat's coat, wrap him in a towel to prevent grooming.

- Call the veterinary surgery immediately.
- Provide as much information as possible, including symptoms, substance ingested, quantity, time elapsed since ingestion. It is always useful to retain any packets that may have contained the substance.
- Follow the vet's advice regarding removal of the contaminant (if appropriate) or any first-aid measures before taking your cat, in a secure basket, to the surgery.
- Ideally, travel with a companion to ensure your cat does not groom during the journey.

Common poisons

There are many substances within the home that are potentially poisonous to cats, including:

Household products

- human medicines such as aspirin, paracetamol, non-steroidal anti-inflammatories, laxatives, antidepressants and over-the-counter medicines such as cold remedies, antihistamines, vitamin and mineral supplements
- cleaning products such as bleach, cleaning fluids and creams, deodorants, deodorizers, disinfectants (particularly phenolic compounds such as Dettol which turn milky in water), coal-tar shampoo, furniture and metal polishes, fabric-softener sheets, dishwashing detergent
- motoring products such as antifreeze, brake fluid, petrol and windscreen-washer fluid
- beauty products such as hair dyes, nail polish and remover and suntan lotion and sun block with zinc

- decorating materials such as paint, paint remover, white spirit and wood preservatives
- miscellaneous household items such as mothballs, pot-pourri oils, citrus-oil extract, photographic developer, batteries, pennies (post 1982) and shoe polish

Pesticides

- insecticides (insect killers) such as organophosphates and pyrethroids
- molluscicides (slug and snail killers) such as metaldehyde and methiocarb
- fungicides (for treating fungal infections: mildew, rose black spot etc.) such as thiophanage-methyl and benomyl
- rodenticides (rat and mouse killers) such as brodifacoum, alphachloralose, difenacoum, chlorphacione, cholecal-ciferol, bromethalin and coumatetralyl

Plants

- indoor plants including dieffenbachia (Dumb Cane), lilies, chrysanthemum, ivy, hyacinth, cyclamen, ferns, poinsettia
- outdoor plants/shrubs including azalea, iris, lily of the valley, marigold, cornflower, daffodil (bulb), poppy, box

The following substances are currently the most common causes of poisoning:

Permethrin There has been a great deal of publicity recently after a number of cats died or became seriously ill after coming into contact with flea and tick preparations manufactured specifically for use on dogs (see Chapter 3 for further details).

If you want to give your cat hours of fun, leave a paper bag or cardboard box out for him to explore.

Slow blink at a cat to make friends; if he slow blinks back it's a very good sign.

Locate your cat's food as far from the cat flap as possible otherwise you may end up feeding your neighbours' cats too!

Cats feel secure in high places so always provide tall perches indoors for your cat; a wardrobe or shelf will do nicely.

Keep your cat's water separate from his food as this will encourage him to drink more.

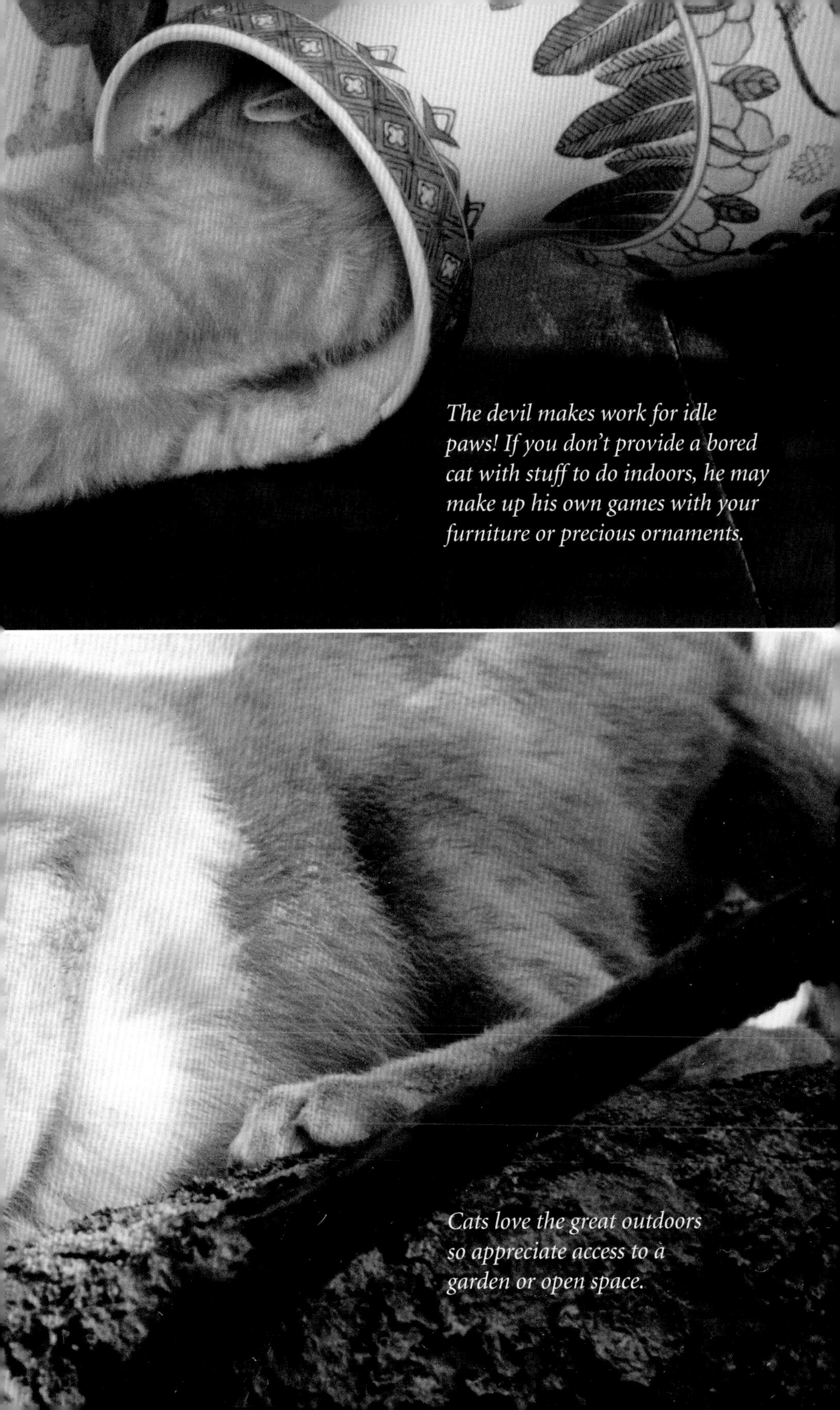

The devil makes work for idle paws! If you don't provide a bored cat with stuff to do indoors, he may make up his own games with your furniture or precious ornaments.

Cats love the great outdoors so appreciate access to a garden or open space.

If your cat starts thrashing his tail from side to side when you are stroking him, stop straight away. He probably isn't enjoying it nearly as much as you are.

The VPIS (Veterinary Poisons Information Service) has been directly involved in eighty-one cases in which the cat died. Permethrin is used in flea and tick collars, sprays, shampoos and 'spot' treatments (liquid deposited on the skin on the back of the neck). Cats are particularly susceptible to permethrin toxicity; even having contact with a recently treated dog could result in symptoms. Permethrin is also a constituent of some wood preservatives and insecticides.

Common symptoms

- tremors
- skin twitches, ear flicking
- drooling
- depression
- loss of appetite
- vomiting and diarrhoea
- breathing difficulties
- disorientation
- vocalization
- seizures
- rapid heart rate

Prognosis There are no specific antidotes but the prognosis is good providing prompt veterinary intervention takes place.

Lilies Easter lilies, tiger lilies and various others are toxic to cats, causing kidney failure and death. These flowers have become increasingly popular over the past ten years and they often form an integral part of bouquets and indoor flower arrangements. The VPIS has been involved in twenty-six cases in which the cats have died or been put to sleep as a result of ingestion. All parts of the lily are poisonous; one leaf can be enough.

Common symptoms

- vomiting
- depression
- renal failure

Prognosis The prognosis is good, if the diagnosis and treatment occurs rapidly; otherwise it's very poor.

Paracetamol Unfortunately some owners administer paracetamol to cats in the mistaken belief that it will be a safe pain reliever. On the contrary, one tablet is enough to kill a cat. The VPIS has been involved in a total of twenty-four cases in which the cats died or were euthanased as a result.

Common symptoms

- vomiting
- depression
- swollen paws
- breathing difficulties

Prognosis The prognosis is good, providing there is prompt veterinary intervention and the appropriate antidote is used – e.g., N-acetyl cysteine or ascorbic acid.

White spirit/Petroleum distillates These substances usually affect a cat through coat contamination, as he will immediately endeavour to remove it by grooming and therefore ingest it. Fuel spills in the garage or on the drive and paint brushes soaking in white spirit are common sources of contamination.

Common symptoms

- irritation and ulceration of the mouth
- vomiting
- anorexia
- skin irritation
- breathing difficulties
- fluid on the lungs

Prognosis The prognosis is good, providing the cat is decontaminated as soon as possible and veterinary support (including ventilation if necessary) is provided.

Disinfectants The most common offenders are phenol compounds (those that go 'milky' when added to water). The cat becomes contaminated when the disinfectant comes into contact with his fur and the damage occurs as he tries to clean himself.

Common symptoms

- irritation and ulceration in the mouth
- vomiting
- salivation
- skin irritation
- loss of appetite

Prognosis The prognosis is reasonable if supportive measures are taken quickly.

Antifreeze (ethylene glycol) Antifreeze products usually contain one of three active ingredients: ethylene glycol, propylene glycol or methanol. The ethylene glycol is by far the most toxic: one half tablespoon is lethal to a 4.5kg cat. The VPIS has been involved in thirty-seven cases in which the cats died as a result.

The ethylene glycol has a sweet taste that can be attractive to animals. It is readily available in most domestic garages as it is used routinely in cars.

Common symptoms

- vomiting, increased urination and drinking, hyper-excitability, seizures
- increased heart and respiration rate
- severe depression, dehydration, vomiting, diarrhoea, kidney failure, death

These usually appear in stages over a period of up to three days after the antifreeze was ingested.

Prognosis The prognosis is very poor, unfortunately, unless veterinary treatment is started within the hour (the specific antidote is ethanol); even then it's guarded.

Low-toxicity household items

While the consumption of any non-nutritional substance should be actively discouraged, there are some common household items that are less toxic than others and shouldn't cause major problems. However, never take any chances and call your vet straight away for reassurance if you think your cat may have consumed any of the following:

- antacid tablets (over-the-counter)
- artificial sweeteners
- Blu-Tack
- chalk
- charcoal
- coal
- expanded polythene
- folic acid

- matches
- oral contraceptives and HRT tablets
- silica gel
- wax candles/crayons

Prevention is better than cure

As with most things, it is better to prevent it happening than to cure it. Taking some practical measures in the home will greatly reduce the risk of your cat getting access to anything potentially harmful.

- Keep all human and veterinary medicines separate in locked cupboards.

- Always read labels carefully to ensure that all products used in the house or garden are safe for pets.

- Always follow the manufacturer's instructions when using any pesticides, herbicides or fertilizers.

- Store products for the garden in a secure container inaccessible to your cat.

- Always administer veterinary medicines to your cat according to the vet's instructions and for the recommended period of time.

- Always ensure floors and surfaces are dry after using any detergent or disinfectant before allowing access to your cat.

- Keep sewing boxes out of reach; needles and threads can cause serious problems if they are ingested.

- Read the instructions on any product for your cat before use. Never use a product specified for dogs only.

- Clean up any antifreeze spills immediately and dispose of any antifreeze-contaminated rags or paper towels in a sealed container.

- Regularly check your vehicle for antifreeze leaks.

- Store antifreeze out of the reach of your cat.

- Use antifreeze products that do not contain ethylene glycol.

Safe plants
On a positive note, here are some plants that you can safely have in your home and garden. This is not an exhaustive list but it should include some of your favourites:

Garden plants
- bamboo
- buddleia
- camellia
- hollyhocks
- lavender
- roses
- snapdragons

House plants
- African violet
- begonia
- spider plant

Indoor/outdoor herbs

- basil
- catmint
- chamomile
- coriander
- cress
- dill
- lovage
- mint
- parsley
- rosemary
- spearmint
- tarragon
- thyme

Cut flowers

- roses, roses, roses (my favourite's yellow if anyone's asking!)

Securing the garden

Even if you adopt all the recommendations made previously in this chapter, for your cat there is no substitute for the wind in your fur and a bit of leaf-chasing. If you have a garden, there is a possibility that it may be suitable for securing to keep your cat safe and the neighbour's cats out.

Benefits of securing a garden

- enables cats previously kept indoors to enjoy the benefits of a safe place to explore outdoors
- more natural indoor/outdoor lifestyle

- allows access to the garden without need for strict supervision
- prevents the risk of road accidents
- prevents fighting with other cats in the territory
- removes the risk of your cat being a nuisance to neighbours by toileting in their gardens
- avoids the risk of injury from dogs
- reduces the risk of theft

There are two options available to you if you decide that you want your cat to get the benefit of the great outdoors. The first option is to secure your whole garden so that your cat cannot get out and the second is to erect an enclosure that contains your cat in a section of your garden. It would be ideal to enclose the whole garden, but most systems require six-foot fencing to be installed along the entire perimeter and this isn't always feasible. Trees and sheds are also an issue as many are positioned in close proximity to boundary fences and this can provide an escape route for cats.

Inverted bracket and wire system

This is installed using a pre-existing six-foot fence, utilizing the timber fence posts. Metal angle brackets are attached to the top of the fence posts and inverted into the garden space by about 2 feet at a 45-degree angle. Stretcher wires are then run between the metal brackets to support a strong wire mesh (size no more than 1 inch) or netting.

The disadvantage of this system is that it doesn't prevent determined cats from coming into the garden from outside. Sadly, once in, it's tough for interlopers to get out and that could be dangerous for you and your cat if the poor things panic. The only way to guarantee this doesn't happen is to obtain the full cooperation of all your neighbours (either side and at the bottom of your garden) to produce a similar system their side of the fence (so that the eventual structure looks Y-shaped). In my experience, the chances are slim of getting everyone to agree to this.

Trees or sheds in close proximity to the boundary, as mentioned before, will make this system ineffective. Branches can be cut from the tree and a collar fixed round the trunk, extending from it to the bottom of the lowest branch. If erected appropriately it would stop cats climbing to the lower branches, thereby denying them access to the tree. However, it can still potentially be used by other cats outside the garden to gain access.

If your garden has a complicated mix of walls, fences and heights, professional installers could be asked to assess your garden and, if it's considered suitable, fix the system for you.

Metal fencing

This may not be everyone's cup of tea but high tensile steel fencing with a painted finish is available. It doesn't look as bad as it sounds and it doesn't warp and is virtually maintenance-free. Cats climb fencing by scaling it, rather than leaping straight to the top, and a metal finish would make it impossible for them to get the necessary grip. Unfortunately, planning permission and neighbours' consent may be an issue with this system but it's worth investigating. Cats will also need alternative outdoor scratching facilities as fence posts and fences are often used for territorial marking and claw maintenance.

Wooden pole or PVC drainpipe system

This system works by fixing either one wooden pole or a series on the inside of your fence near to the top. These poles are fixed to rotate or spin if they are pulled, so a cat would, in theory, attempt to grip to get to the top of the fence and then fall back as the pole rotates. A twin-pole system will keep even the most determined cat within the garden.

The benefit of this system is that your neighbours can't complain, as it's invisible from their side of the fence. The disadvantage is that other cats can come in but, once again, not get out. You could fit the system to the top of your fence (neighbours permitting), thereby preventing other cats from coming in.

This system does require the erection of six-foot fencing (or walls) round the whole perimeter. However, once installed it does have some additional benefits as the manufacturers maintain that gates, trees, window ledges and roofs can be protected with the poles.

This type of system can also be installed using a specific type of connector and PVC draining pipe.

Free-standing galvanized post and mesh

This system is ideal for those properties without adequate fencing. It can easily be installed by any enthusiastic DIYer and, once again, the inverted top and flexible mesh netting deter even the most proficient climber from escaping. They are designed to be fitted inside the boundary of your property, well enough away from any existing walls or fences to avoid the risk of the cat using them to gain some purchase on the netting.

A few words of caution . . . Some trees have preservation orders, so be sure to check with your local council before hacking down trees that prevent you from securing your garden. To avoid any unnecessary bad feeling between you and your neighbours, always check with them, your solicitor and the council's planning officer before erecting any new fencing.

Any gates that give access to the garden will have to have similar treatment too to ensure the area really is cat-proof. Gaps below the fence, holes in fencing and holes that are dug under fencing by local wildlife all represent possible escape routes for the more Houdini-like cat.

Watch out for squirrels if you opt for a mesh-netting system; they can chomp away and create holes in the mesh that will compromise your cat's security.

Electric system

There has been a great deal of controversy about the use of electric fencing for domestic pets. If you have experience of the type of electric fencing that confines horses and farm animals, you will understand many people's reservations. I use such fencing for Naiad and Rupert, my aged Arabian horse and rather mischievous donkey. They have the utmost respect for it and so do I; the odd time I have inadvertently been zapped by it, I can honestly say, wasn't pleasant.

However, forgetting that type of system for a moment, there is an alternative kind that is a 'radio' rather than an electric fence and is specifically designed for use in gardens to contain cats and dogs. Encircling the boundary, it consists of wire that carries a low-power, low-frequency radio signal from a transmitter. Your cat wears a tiny battery-powered receiver on his collar, which picks up a coded signal from the wire. When he approaches the wire (any distance between 2 feet [60cm] and 15 feet [4.5m], depending on what suits you and your garden) he hears a continuous warning beep from the receiver on his collar. If this means nothing to him and he proceeds further towards the boundary, he will receive a small shock (similar to the static electricity that you occasionally experience when touching a car door). This isn't physically harmful but it will startle him and the theory is that this will stop him – particularly if it happens a couple of times – from advancing or going in that direction again.

I have no personal experience of using this type of system. I have spoken to many owners who have installed it and they are delighted. Their cats don't go near the boundary fences and have safe access to a garden that wouldn't have been possible if the system hadn't been in place. Conversely, though, experts in cat behaviour tell me they are yet to be convinced that this system is as foolproof as suggested. Some believe that cats may be startled and jump forward; the sensation is felt on the cat's neck, so there is no definite association with something in front of the cat. The sensation may also be paired in its brain with another stimulus that may coincidentally occur at the same time, risking possible redirected aggression towards another cat or human.

I am very pragmatic about such things. If the alternative is a life of confinement and the cat is clearly struggling to cope I am sure that, if the garden could not be secured in any other way, I would certainly investigate thoroughly the suitability of an electric fence.

Purpose-built enclosure

Sometimes it is just not possible to secure your entire garden, but this isn't your only option. A cat enclosure outside is not necessarily the ideal but it's better than nothing and will still enable your cat to benefit from all the sights, sounds and smells of the challenging environment outside the four walls of your home.

Cat enclosures can either be attached to the house or conveniently located some distance from it. The benefit of having it attached is that at least one side of the enclosure is already constructed, i.e., your house wall, and it would be possible for your cats to choose when to use it if a cat flap is fitted that gives access to the pen. Often, too, a fence or wall between properties can be utilized as one side of the enclosure.

If the enclosure is situated, for example, at the bottom of your garden you will need to take your cat there in a secure basket to prevent any risk of him jumping out of your arms and over the fence. This means that you will have to be in charge of when your cat gets his fresh air and it may not always coincide with his wishes. It does also mean that your cat may become exceedingly frustrated if every time he sees or smells something interesting he is taken back indoors. However, if the construction is secure enough with all the necessary shelter and warmth, there is no reason why your cat can't remain there all day.

The ideal construction of a typical enclosure is strong wire mesh on a wooden frame (4in × 2in) with an enclosed roof of either further wire mesh or PVC corrugated sheeting with a UV filter on a wooden frame. The latter is the best option if the enclosure is designed to offer all-year-round protection against the elements. This roof should ideally be slightly sloping for drainage purposes. The height of the structure should be at least 6 feet, to allow easy access to humans and enable the cat to climb up high.

If the internal dimensions allow, shelter can be provided inside by a small summerhouse or garden shed, insulated and lined with a washable surface to make it easy to clean. This can house any dry food that is left out for your cat (not wet food, as it may attract flies in warmer weather) and a warm bed for him to curl up in. You might also, ideally, install an outdoor electric supply in the shed to enable the use of a heated pad for the more fine-coated individual!

Wooden platforms and shelves should be attached to the

frame of the enclosure to enable the area to be used to its maximum three-dimensional potential. The area should also contain a water bowl (or water feature), a stack of wood (great for encouraging insects and fun to climb), pots of grass or catmint, a covered litter tray and anything else that you feel your cat may enjoy.

Neighbours' cats can pose a problem as they are naturally attracted to the sight of other cats enclosed behind wire fences and can often be found taunting them mercilessly. Aim to make it difficult for all but the most determined cats to gain access to the roof of the enclosure you construct, since having a strange cat stomping repeatedly overhead may well put off your own cat from using it. Positioning plants and pots outside the enclosure in sufficient depth will provide your cat with that all-important camouflage when he is using his toilet and also deter cats from getting up close and personal to the pen at ground level.

Security is important, too, if you are considering leaving your cat in the enclosure unattended. You will need to construct an access door in order to clean and generally maintain the pen. This should be securely locked and bolted with a strong padlock.

A couple of points:

- Consult your local planning department to check on any regulations with regard to garden enclosures.
- Ensure that the wood preservative used is safe for cats.

Cat flaps

If you make the decision to let your cat go outside, there are further considerations regarding his mode of entry and exit. You have a choice:

1 Shut him out during the day, if you are going out to work
Advantages
- You can maintain the security of your home as no other cats can come in.

Disadvantages
- Your cat may not want to be out all day.
- You will need to provide shelter outside against inclement weather.

2 Let him out, on demand, under supervision when you are at home
Advantages
- You always know where he is and he soon gets used to you being instrumental in his excursions outdoors.
- Your home is secure from invasion providing you shut the door when he is out and don't let any old stray wander in.

Disadvantages
- It limits his freedom and the amount of exercise he gets.

3 Fit a cat flap and leave it open twenty-four hours a day
Advantages
- It gives him freedom to decide when he goes out and for how long.

Disadvantages
- Unless your cat flap is exclusive entry any cat could come in.
- Cat flaps attract cats, whether you like it or not!
- Your cat is statistically at more risk of injury at night.

4 Fit a cat flap and leave it open during the day but shut him in at night
Advantages
- Cats often adapt well to coming in at night (it's usually warmer indoors).

Disadvantages

- Sometimes he doesn't want to come in at night (especially in the summer) and you have then to decide whether to stay up waiting for him (not recommended) or leave the flap open.
- The cat flap will still attract other cats.

Road traffic accidents

A survey conducted by one of the UK's leading pet insurers revealed that a cat is run over on the road every two and a half minutes. Nearly a quarter million cats were involved in road traffic accidents in 2005/6, half of them between the ages of seven months and two years. Cats can develop good road sense as they get older but many youngsters are focused on reaching interesting things on the other side of the road and the traffic is almost irrelevant. Many road accidents happen in poor light at dawn, dusk or night-time. Therefore, statistically, your cat would be safer if he was kept in during the hours of darkness in built-up areas.

The cat-friendly garden

Whether your cat is confined in it or free to roam, it's important to ensure that your garden represents a haven of security, fun and excitement. Thoughtful garden design can be pleasing for you and your cat; here are some tips:

- Have some different levels in your garden and opportunities to climb, including trees and sheds. If your garden is small and cannot accommodate such things, then consider fixing a series of small shelves to the back wall of the house where your cat can sit and survey his territory.

- Provide your cat with his very own outdoor toilet. Choose an area comparatively close to the house if he is a little insecure about going outside. Make sure any chosen area is against a wall, fence or dense shrubbery with plants either side to provide privacy and security. Dig an area of soil (approximately 24 in [60cm] × 18 in [45cm]) to a depth of 2 feet [60cm] and fill two-thirds of the depth of the hole with shingle for drainage. Fill to the top with fine sand (children's sandpit quality). Faeces can then be removed with a litter tray scoop and the sand topped up. This will not freeze or become waterlogged and a layer of peat or soil can be sprinkled over the top to disguise its use.
- Avoid vast open areas of lawn, patio or decking that provide no camouflage for your cat when he is wanting to 'recce' the area without being seen.
- Position a selection of pots and tubs near the exit and entry point to the house – e.g., near the cat flap or door. This will provide immediate protection for your cat, allowing him to check things out first before moving away from the safety of the home. Choose a variety of shapes, including tall and broad ones, but make sure you put stones over the top of the soil to avoid your cat using them as a convenient toilet.
- Stock your borders with a variety of plants and dense shrubbery to provide your cat with private areas and protection from extreme weather.
- Leave a weatherproof container outside to collect rainwater. If you really want to be extravagant, consider an outdoor water feature – your cat will love it!
- Protect any fishponds with cat-proof netting to avoid any watery accidents.
- Don't remove all your grass in favour of a low-maintenance

garden. Cats love to eat grass, so a small area should be provided. Growing grasses in a pot is a compromise if you are desperate to abandon the lawnmower. Don't, however, be tempted to plant the type of ornamental grasses with barbed leaves, as these can get lodged in your cat's throat or nose and cause problems.

- Steeply banked gardens can be intimidating for your cat as, when he's indoors, strange cats will be seen as high up and make him feel vulnerable. Covering the lower part of any floor-length windows or glass doors will help, as will positioning a high scratching post (cat activity tower) or piece of furniture near the door to give him a better vantage point.
- If your garden is south-facing it's important to include areas that are designed to provide shade for your cat in the hot weather. Cats with white fur are susceptible to sunburn on their ears and noses and can develop serious skin conditions as a result. My own cat Mangus has issues with direct sunlight as her coat is sparse in places, so she makes good use of shelter in the garden, including the patio umbrella.
- Grow catnip! This is best introduced as an established plant but be prepared to have it flattened as your cat rolls on it or chews it down to woody stalks.
- Your cat will want to scratch in your garden, so make sure posts are available to avoid damage to precious trees.
- If you feel adventurous, construct a wooden climbing frame for your cat to enjoy as his outside play area.
- Ensure that all wood preservatives used are cat-friendly.
- Ensure all products used in the garden are safe for your cat and follow the manufacturers' instructions carefully.
- If you are feeding birds in your garden make sure that you have a bird table positioned on a high pole, away from fences and trees, to prevent your cat from gaining access.

This next section isn't usually the sort of thing you expect to find in a book extolling the virtues of cats, but, whether we like it or not, cats can be a problem to other people. Often disgruntled neighbours, objecting to the use of their garden as a popular cat latrine, fight a continuing battle with the cat owners. Frustration can motivate them to weird and wonderful lengths to remove your cats from their prize gardens, so here are some suggestions for preventative measures you can offer should the situation arise. As a bonus, you will be perceived as a caring and conscientious pillar of the community!

Cat deterrents

First of all, make sure the cat is the true culprit (often fox and hedgehog faeces are confused for the feline variety).

- 'Shoo' the cat away by shouting or clapping your hands.
- Squirt water at the cat.
- Stop feeding birds and squirrels with food scraps; this can attract cats.
- Install an automatic garden spray with an infra-red detector that is activated by movement in the garden (this avoids the need to stand guard!).
- Grow ground-cover plants or protect recently turned-over soil with netting.
- Water flowerbeds last thing at night if the toileting takes place in the early hours.
- Citrus peel, cayenne pepper, oil of peppermint and eucalyptus can deter cats.
- Better still, get a cat of your own and then other cats will stay away. (Of course, your cat will probably poo in your garden, but what the heck!)

Hunting

I often think that owners are in a persistent state of denial about what a cat actually is. Whether we like it or not, a cat is (and always will be) a top-of-the-food-chain predator. It is an obligate carnivore and it will, if given half a chance, kill things. This is not for pleasure or even just to annoy you. It is a primitive drive that has very little connection with hunger. Appetite will merely dictate how hard the cat tries and whether or not it kills or eats anything it catches. Even my Mangus, the least cat-like cat you could imagine in many respects, indulges in the ancient art of 'frog flipping' (brown side, green side, brown side . . .). I don't encourage it and she easily tires if I gently remove the frog and send it on its way.

I honestly believe we should accept prey-catching as a fact of life when living with a cat, but if you really can't bear it (and who wants the local wildlife to end up dead on the kitchen floor just because of his 'primitive drive'?) there are measures that can be taken to limit the loss of life.

- Don't actively encourage birds and wildlife to the garden by having bird tables and putting out food for them.
- If your cat is a prolific hunter at certain times of day, then limit his access outside during these hours and compensate with lively predatory games.
- Put bells on a safety collar or use a commercially available sonic bleeper deterrent so that the prey gets a little warning (these have been shown statistically to reduce the amount of prey caught).
- Shut inner doors to keep your cat in the kitchen at night if he has access outdoors. Any little 'gifts' will at least be contained in one room and not on the carpet by your bed, ready for your bare feet to find them in the morning.

Missing cats

One of the most common reasons for owners keeping their cats indoors is the fear that they may go missing. I have 'lost' Annie, Lucy and Bink (three of my much-loved cats) at some stage in their lives; all three returned. Admittedly Annie was lost for only twelve hours, but that particular incident taught me a huge lesson: *Some cats just don't want to be found – they'll be in when they are good and ready!* I still don't know where Lucy and Bink disappeared to; they both went missing for over six weeks. I did everything that I am about to recommend that you do should your cat go missing, but they both just wandered in one day as if nothing had happened.

There are obviously some precautions you can take to limit the chances of your cat going missing. Neutering at six months of age will reduce the distance that your cat is likely to roam, and microchipping (see Chapter 3) will ensure that, when he's found, you will be contacted. It is important to remember to keep the microchip database – e.g., Petlog – updated with any new contact details or address.

Fitting an appropriate safety collar with your telephone number on it makes for a quick form of identification, but don't rely on this alone as collars are easily caught on branches and lost.

When the fateful day comes and your cat doesn't return for his tea, it is the easiest thing in the world to go into a flat spin. I speak to loads of owners who call my office in a panic to say their cats have gone missing. They are so distressed they literally cannot gather their thoughts to decide what action to take. To avoid this ever happening to you, make a mental note that this section is here. If your cat doesn't come home one day, just turn to the next page and follow the instructions carefully.

OPERATION LOST CAT!

Action plan

Step 1

Don't panic. There may be something absolutely fascinating going on outside that overrides his normal desire for food at the moment. He may even be planning alfresco dining tonight with something furry on the menu. Always give him the benefit of the doubt and wait until morning to take further action, even if you haven't seen him since breakfast. Think about whether there may be a seasonal pattern to your cat's vanishing act; did he do the same sort of thing this time last year? Some cats just like to have a bit of a wander and it is entirely likely that is what he is doing now.

Step 2

If morning comes and he is still not home, then the next step is to presume that:

1 He's been in the house all the time and found a new secret place; or

2 He's been a little too nosy and got himself shut in somewhere nearby.

So first check every room in the house, including chimneys, and lofts or cellars, cupboards, under curtains, behind appliances in the kitchen, under chests of drawers, etc. Bear in mind that if there is a gap big enough for him to squeeze his head into, then the rest of the body can easily follow. I lost my Spooky at least a dozen times in the house; to this day I still don't know where she holed up.

Once you have exhausted all the options in the house search, you need to explore further afield. First check your own garden (front and back), garage, shed and any other outbuildings.

Look in dustbins, water butts, compost bins, under hedges and shrubs. While searching take time to stop and listen for scraping sounds or faint miaows.

Still not home? Ask your neighbours to check their garages, sheds and greenhouses. Make the request of neighbours either side, across the road and in properties behind your garden. Think about where you have seen your cat or the neighbours who have commented that your cat has been in their garden and target them particularly.

Step 3

Another day has gone by and still he hasn't returned. Now is the time to make it official that your cat has gone missing. Find a recent clear photograph and make a note of all the details that may be required – for example, his name, age and colour, short- or longhaired, male or female. Does he have any distinguishing features? Was he wearing a collar? Is he microchipped? If the answer to the last question is 'yes', then contact your microchip company first to register him missing. Once you have done this, call:

- local vets
- police station
- local animal-rescue charities
- local newspaper (to put an ad in the 'Lost and Found' section)
- local boarding catteries
- local radio
- RSPCA
- local council (the Environmental Health Department will know if any cats have been found on the road – sorry, but you have to know)

Step 4

You can go to Step 4 on the second day if you are particularly keen to do everything as soon as possible; otherwise you may want to wait until the third day. Now is the time to get out there and walk around, but don't go alone after dark. Don't forget to take a torch and your cat's favourite pack of biscuits with you (many cats will answer the shake of a biscuit box before their name). Investigate garages, buildings and empty properties (watch out you don't trespass, though). Stop regularly to call out his name and then wait quietly to see if you can hear anything.

Step 5

Day three or day four, it's now time to let as many people as possible know he is out there somewhere. Make a leaflet (easy to do if you have access to a computer) and include a large photograph. Offer a 'substantial reward' for his return but don't specify the amount. Just include your telephone number as a contact and don't give away too much information about your cat. Now distribute these leaflets as follows:

- local shops
- post office
- doctor's surgery
- local supermarket
- veterinary clinics
- village hall noticeboard
- local school
- every letterbox in the area
- cover some in clear plastic folders and attach them to telegraph poles and lamp posts

You should also make another phone call to the local vet's to see if any cats have been found and check noticeboards in your local shops and your local paper for any 'found cat' notices.

Step 6

If someone calls to say they have found your cat, don't forget to ask all the important questions to make sure they can identify distinguishing features and confirm it really is him. Arrange to meet them and take with you your cat's basket, but don't go alone. If it is your cat, make sure you call all the relevant agencies and inform them that you have found him. You will also need to remove posters, etc.

Step 7

When you get him home give him a little of his favourite food and have a quick check to see if there are any obvious signs of injury. If he seems well, then don't worry, just be glad he is home. If you have any doubts about his health, a consultation with the vet will put your mind at rest.

Step 8

If the previous two steps just haven't happened for you, don't despair. Keep telling everyone he is still missing and keep looking. Remember Lucy and Bink, who returned after many weeks, safe and sound and entirely of their own accord! Some cats take even longer, so if you don't have evidence that he has come to harm there is always hope. Whatever happens, don't worry that he is out there somewhere hungry and alone; cats are extraordinarily resourceful and I can assure you he will be coping a lot better than you may think.

Many pet-insurance companies will cover the costs relating to the search for your missing cat. Some give free membership with 24-hour 'Lost and Found' services.

When they just don't come back

Sadly, there will always be some cats that go missing and never return. Somehow this can be the most difficult form of loss; you have no focus for your emotions and no closure in the form of a burial or cremation. No matter how many months go by there is always a faint hope that he may, one day, return. Eventually, unfortunately, you have to let go.

Under these circumstances owners often grieve for longer for their beloved cats; this is perfectly understandable. Chapter 9 may provide some reassurance that the feelings you are experiencing, if you find yourself in this situation, are normal.

Cats returning to their previous homes

Occasionally when cats go missing they have a very definite task in mind. Some cats are more attached to their territory than to their owner, so if they move within a relatively short distance of their familiar hunting ground they will probably try to return. This journey may traverse busy roads, railway lines and even rivers, but the drive to go back is extremely strong. If you have moved to a new address within a couple of miles of your previous home and your cat goes missing, if he was a great hunter in your last place that's where you should start looking in the first instance.

- Make sure your cat is microchipped and your address change is notified immediately.
- Keep your cat indoors for a couple of weeks if possible (some cats really object) and feed small tasty meals three or four times a day.
- Allow your cat outside first just before he is due a meal.
- If he goes missing, inform your previous neighbours and ask them to look out for him.

- Inform the new occupier that your cat may be on his way back.
- Ask everyone not to feed your cat but to notify you immediately he is seen.
- If he is easy to handle, ask a neighbour to pick him up and secure him somewhere to await your arrival.
- Take him home and provide a tasty meal and plenty of predatory games with toys.

I once gave advice to a lady who persevered with her cat, collecting and bringing him back two or three times a week. After six months he finally got the message. Some owners, however, particularly if their cat has to travel over dangerous roads to get there, encourage a neighbour or the new occupier to take him on and allow him to live out his days in the area he loves.

The law and your cat: disputes with neighbours

There is a great deal of legislation in place to protect our pets. The recent addition of the Animal Welfare Act 2006 (effective from 6 April 2007) is a positive step forward, as it expands on the current laws against cruelty to animals by focusing on the welfare of our pets. It makes owners and keepers of animals responsible for specific welfare needs, including:

- a suitable environment
- a suitable diet
- the ability to exhibit normal behaviour patterns
- to be housed with, or apart from, other animals (depending on the species)
- to be protected from pain, injury, suffering and disease

As part of the Animal Welfare Act, a code of practice is

being introduced for most companion animals to give owners guidelines and advice about how to care for their pets appropriately. There are specific sections of the Act relating to the sale of cats through pet shops and fairs. Cruelty and irresponsible pet ownership is still a major issue in the UK and any legislation that assists in educating the general public and preventing or stopping these problems is a good thing.

It is very likely that you will never be in a position requiring you to have knowledge of the law relating to your cat. There is only one ever-increasing situation in which it would be useful to know where you stand. I am speaking, of course, of the alarming rise in disputes between neighbours concerning the behaviour of a pet cat.

In some areas they are seen as Public Enemy No. 1, with damage to plants in the garden, vehicles in the driveway and property within the house itself. The main matters of concern are:

- cats using other neighbours' gardens as toilets and digging up tender plants
- cats entering houses to eat another cat's food
- cats breaking cat flaps to gain access to neighbours' houses
- cats fighting with other cats in their own homes, causing injuries requiring veterinary treatment
- cats attacking owners attempting to separate fighting cats
- cats spraying urine, urinating or defecating in neighbours' houses
- cats causing other damage in neighbours' houses
- cats sleeping on car bonnets or roofs and scratching paintwork

Unfortunately we are a nation of cat lovers and cat haters.

Whenever I do radio interviews that include a phone-in there is always at least one 'cat hater' who asks for advice on keeping the 'wretched things' out of his (or her) garden. I think we just have to accept that we are not all made the same and for some inexplicable reason not everyone 'gets' cats the way we do!

Disputes with neighbours can be extremely distressing; there is often a huge divide between your opinion and that of your neighbour and arbitration is virtually impossible. The main thrust of most arguments is that you think your cat is doing what cats do and your neighbour thinks it doesn't want your cat to 'do what cats do' in his house, therefore you have to get rid of him. I really hate these cases because there are such intense emotions on both sides and I can honestly understand how each party feels, but people have to be realistic. We live cheek by jowl with many other people on this overcrowded island. We therefore also live in close proximity to their pets and we have to develop a degree of tolerance, providing that the pet owners are doing everything reasonable to prevent them from being a pest. The law is quite specific about dogs and the need to prevent roaming or loss of control in a public place. Cats are categorized quite differently; the laws relating to trespass do not apply to cats. However, this doesn't give us *carte blanche* to chuck our furry friend out in the morning and take no responsibility for what he gets up to when we're at work.

Here are some suggestions, tried and tested, that may help find some common ground in disputes with neighbours. The following basic recommendations apply to any disagreement that requires discussion:

- Arrange to meet on neutral territory or, if appropriate, invite your neighbours round to discuss the problem.

- Plan your strategy beforehand and have notes prepared as you don't want to forget anything should things get heated.
- Understand that the best compromise is to meet halfway, so don't get too worked up about the rights and wrongs; be prepared to offer something to resolve the problem.
- Don't apportion blame or admit liability; this is counter-productive.
- Confirm everything in writing afterwards to ensure you both understand what was agreed and put time limits on any actions that need implementing.
- Keep a log of any phone calls to or from the neighbour, particularly if threats are being made.
- Remember, whatever happens, you have to live next door to each other.

If, for example, your cat is entering your neighbour's house and attacking the family cat:

- If your neighbour has a cat flap that isn't exclusive entry, offer to replace it with one that will allow only the household cats in or out. If he or she is reluctant to introduce facilitating cat collars, then recommend a cat flap that reads the resident cats' microchip.
- Your cat will probably stop going to the house if the neighbour's cat flap is temporarily boarded up for a period of time. Your neighbours will have to provide an indoor litter tray and let their cat out under supervision for a while but the little victim will probably be grateful for the moral support!
- If your cat is getting in through open windows and doors, offer to pay towards 'fly screens' or grills that can be put in place to allow air to circulate but prevent anything coming into the house.

- Advise your neighbour not to intervene directly in fights. Recommend a blanket, cushion or pillow to separate the cats if intervention is necessary.

These cases hardly ever get to court; there are only a handful of relevant precedents under the Animals Act 1971. Fortunately there isn't really a case to answer, as your neighbour has a duty to protect his or her property and it could be argued that the provision of a cat flap or even an open door constitutes a 'voluntary acceptance' of the risk of another cat coming in. Cats are territorial creatures; if your neighbour's cat cannot defend his territory, that is unfortunate. The invading cat is not 'aggressive' or 'dangerous'; it is just behaving like a cat.

The gardener's grievance is a valid one; it must be difficult to tolerate your plants being regularly uprooted by a toileting tomcat. Show you are a responsible cat owner and do two important things:

1 Provide indoor litter facilities for your cat. Many cats prefer the security of an indoor toilet.

2 Provide a safe latrine area in your own garden (see the section in this chapter on the cat-friendly garden).

If you are making every effort to reduce the nuisance, you will be seen as the perfect neighbour!

Choosing a cattery

There are a number of lifestyle choices that you will need to make that go beyond the indoor/outdoor debate. What happens when you go on holiday or move house?

Everybody deserves a holiday once in a while but many people refuse to go because they are so concerned about leaving the cat. It really doesn't have to be such a problem. If you find a good cattery, and book it sufficiently in advance, you will have a great holiday safe in the knowledge that your cat is receiving five-star care. Your cat won't be at home or in familiar surroundings but, if the cattery is chosen well, he may even be having a little holiday of his own.

There are some very good boarding catteries out there but, unfortunately, there are some dreadful ones too. The Feline Advisory Bureau inspect catteries and provide a listing of all those that have achieved the high standards recommended by the FAB regarding construction, management and day-to-day care. However, not all catteries apply for FAB listing and there are good ones that have somehow slipped the net. There is no substitute for doing your own groundwork and investigating the local catteries.

- Make an appointment to view the cattery. If the owner refuses to show you the premises, look elsewhere.
- Try to stick to those establishments that only board cats if your cat isn't used to dogs.
- Check out the surroundings: are they neat and clean?
- Does the proprietor give you a warm welcome? Does he or she seem well informed about cat care and running a cattery?
- All catteries should be licensed by the local authority, so don't be afraid to ask to see their licence.
- Ask the proprietor what provisions are made to quarantine sick animals (some catteries have a separate enclosed isolation pen).
- Ask to see the food-preparation area and the place where the litter trays are cleaned. Ensure there are separate utensils for use with each pen.
- The accommodation for each cat should have a separate

enclosed and insulated sleeping area and an individual exercise run. The pen should be warm, dry and secure, and big enough to accommodate food, water, scratching posts, litter tray, toys and space for the cat to run around. There should be larger pens available to enable cats from the same household to be kept together.

- Indoor catteries should be avoided as they are difficult to ventilate and may promote the spread of airborne disease. An outside space for cats is in any case much more entertaining for them.
- The pens should be heated in the winter and well ventilated in the summer, with plenty of shade.
- Leading to each pen there should be an outer enclosed corridor that is securely locked at all times to provide additional security.
- Between each unit there should be gaps (minimum 2 feet [60cm]) or full-height sneeze barriers to prevent the spread of airborne diseases.
- Cats should have somewhere to hide within the pen and a high shelf for resting.
- A nice view from the outside run would be great to give cats something to look at during their stay.
- The cattery should have no strong odour of cleaning products or faeces or urine.
- The cats in residence when you visit should look alert and interested. The food bowls should be empty or a significant portion eaten, which indicates their appetites are good and all is well.
- A good proprietor will ask for copious information about your cat – name, age, eating habits, likes and dislikes, and especially whether or not he is longhaired. This is good news because it shows the proprietor understands the need

to groom him on a daily basis. He or she will allow you to bring bedding, scratching posts and other things from home to make your cat feel settled.
- You will be requested to provide an up-to-date vaccination certificate and a medical history with full details of any medication or special requirements your cat may have – for example, some older cats find ramps up to sleeping areas a little difficult, so provisions should be made to ease the transition to the sleeping quarters.

If you are satisfied on all counts, then this is a good cattery! It also means that it will be very busy, so book a long time in advance and be prepared to plan many months ahead for future holidays. You may even want to try a long weekend to start with, to ensure your cat is as happy with the establishment as you are.

House-sitters

A cat's territory is extremely important to him and there will always be those cats that would much rather remain at home when you go off on holiday. If your cat is used to coming and going throughout the day, and cuddling up with you when the mood takes him, it would be perfectly acceptable to arrange for a neighbour, friend or member of the family to pop in twice a day to provide food and check that all is well. If you have an indoor litter tray, if you are lucky, they will service this too!

If you can't find anyone to help out when you go away you may want to consider using a professional house-sitting service. A number of companies exist that provide pet care, plant-watering and general home security when you are away. They usually employ retired professional people who enjoy the chance to experience new places and look after a variety of animals. They can either stay in your home or visit twice a day. These people are thoroughly vetted and come with references; you have the opportunity to interview them prior to booking to make sure they are the kind of people you trust with your cherished cat (and home).

I do have one word of warning if you are considering leaving your cat at home with a sitter, particularly if he is an indoor cat. House cats have a very strong sense of territory and that is the confines of the four walls of your property. You are a very important part of that and your presence may be a great source of reassurance and security. If you go away, there is a tendency for some of the more sensitive cats to feel vulnerable and 'home alone' to ward off any enemy attacks that may occur in your absence. I have even heard of cats taking their responsibility to defend territory so seriously that they launched severe attacks on their cat-sitters and the latter have ended up chucking food through the letterbox rather than risk receiving another wound. Even if your cat doesn't respond in this way he can still become greatly stressed and you may return to find he has deposited urine and faeces around the house in a desperate attempt to mark his territory. If your cat is a sensitive soul, and particularly if he is wary of strangers, it may be better if he has his moments of stress away from the home in a caring cattery.

Moving house

Moving house can be traumatic enough without considering the needs of your cat. However, it is really equally traumatic for your cat to be swept up one day and deposited later in a completely different world. The new home is potentially a scary place where your cat has to establish new territory and familiarize himself with a host of new sights and sounds, all very daunting. If you just can't bear having to worry about your cat as well as everything else on the day of the move, then arrange beforehand for him to have a day at a local cattery, away from all the hustle and bustle.

- On the day of the move, shut your cat or cats into one room that has been cleared of all large furniture. It may be worth using one that contains a fitted cupboard where he can hide if he gets anxious. Leave food, water, bedding, litter tray and other familiar cat toys and scratching posts.
- Make sure the removal team know that the door to the room where he's confined must be kept shut to avoid him going missing at the last moment.
- Once the removal van has gone, your cat can be taken in your car to your new home.
- If you have access to the new property before the removal van arrives, it would be helpful to position a Feliway diffuser (the synthetic feline pheromone plug-in) in a floor-level socket to make your cat feel more comfortable when he arrives. Put it, in the first instance, in the room where you intend to keep your cat for the first few hours while you move in.
- If you have planned ahead you will have asked for the furniture for one bedroom to be stored in the van in such a way that it can be unloaded first. Once this room has all its contents installed, your cat can be left in there with food, water, a bed and a litter tray while you continue to move all the rest of the furniture. He can now be left quietly to explore this new room at his own pace. Another KEEP CLOSED notice on this door will ensure he doesn't disappear.

- Your cat can be allowed into the rest of the house once it is quiet and all the furniture is in place.
- If he is nervous you may want to allow him to explore small areas at a time. Make sure any chimneys or gaps behind cupboards and appliances are blocked to prevent him withdrawing into these inaccessible places. Be relaxed and don't reinforce anxious behaviour by reassuring him.
- If your cat is confident and loves to be outdoors, you may find it difficult to confine him for the recommended period of two to three weeks.
- When you first allow your cat to explore outside, it is best to choose a weekend when you are at home and let him out just before a mealtime.
- Ensure your cat is microchipped and has a safety collar showing your telephone number just in case he roams a little too far in his new territory.
- Your cat may be involved in a bit more fighting than he is used to as he establishes his new territory. Unfortunately this is what cats do, so just keep your fingers crossed that the neighbourhood isn't full of large aggressive felines.

Caring for your cat when travelling

Whenever you travel with your cat you have a legal duty of care, as the law states: 'No person shall transport any animal in a way which causes or is likely to cause injury or unnecessary suffering to that animal.' If you pay a transport company to take your cat on a plane or boat you still have a duty to ensure that your cat is fit enough to travel, as they may refuse to take him if they can't provide the right conditions to ensure his welfare throughout the journey. I have had a number of calls from owners wishing to emigrate and take their elderly cats with them; I think this is one occasion when careful consideration is necessary to ensure this is the right thing for the cat. If your cat is particularly nervous, this may well be one journey too many.

Tips for any journey

- Travelling overnight is recommended as your cat may be used to sleeping at this time. If he is more of a nocturnal creature, then daytime travel would be ideal.
- Feed only a light meal a couple of hours before travel as he will be more comfortable without a full stomach.
- Sedatives are not recommended unless specifically prescribed by your vet. If they are, your cat will need to travel with a certificate declaring what drug was used, the dosage and the time and date of administration.
- Put familiar bedding in the container and introduce your cat to it prior to the journey. Spraying it with synthetic pheromones (Feliway) half an hour before would also help your cat feel more comfortable in strange surroundings.
- Ensure your cat is securely locked in a well-ventilated carrier.
- Secure the carrier to the seat (with the seatbelt) to prevent excessive movement during braking or acceleration.
- Water should be available for offering to your cat at regular intervals during the journey.
- Ensure, if you are travelling by car, that the vehicle is well ventilated.
- If you leave your vehicle unattended at any time during the journey make sure the vehicle is not in direct sunlight and that windows are partially open.
- If you are embarking on a long journey the cat carrier

should be big enough to accommodate a litter tray fixed to its base.

Pet passports and moving abroad

The Pet Travel Scheme (PETS), introduced in February 2000, is the system that allows pet dogs, cats and ferrets from certain countries to enter the UK without quarantine as long as they conform to the rules of the scheme. UK owners can also now take their pets to other EU countries and some non-EU countries and then bring them back without quarantine.

The requirements for a pet passport to enable your cat to travel in this way are as follows:

- Your cat must be microchipped.
- Your cat must be vaccinated against rabies. He must have the complete course as recommended by the manufacturer at least twenty-one days before travelling. Booster vaccinations must then be given at the appropriate intervals. This will enable your cat to enter most other EU countries. If you wish to enter or re-enter the UK, or to enter Malta, Sweden or Ireland (other than direct from the UK), your cat will need a blood test to ensure the vaccine has given sufficient protection against rabies. You will then not be able to re-enter the UK for a further six months.
- Your cat will also require treatment for ticks and tapeworms and you will need the corresponding certificate to prove that this has been done not less than 24 hours and not more than 48 hours before you check in for your journey.

There are a number of complicated requirements imposed by various countries throughout the world for entrance from the UK or for entering the UK from those countries. It is a minefield but the important thing is to prepare yourself with the appropriate knowledge many months before you intend to travel to ensure you don't end up leaving your cat behind. DEFRA (Department for Environment, Food and Rural Affairs) is the best source of all this information in the first instance and, if necessary, they can provide you with contact details regarding travel to or from specific countries outside the EU. They will also give information about transport companies and IATA (International Air Transport Association) approved containers for your cat to travel in, together with the legal requirements necessary for transporting animals by air.

If you are planning to emigrate, take a little time out to think about the change in environment and general conditions that this will mean for your cat. Research the country's attitude to cat care and quality of life and the kinds of new dangers, diseases and parasites that your cat may be exposed to. If carefully thought out, a move abroad can be great for all parties concerned.

CHAPTER 7

One Is Never Enough

MULTI-CAT HOUSEHOLDS AND SOCIAL INTRODUCTIONS

IT IS A WELL-KNOWN FACT THAT I HAVE, IN THE PAST, SHARED my home with as many as seven cats at the same time – and I am not alone, as over a third of all cat-owning households in the UK have more than one cat. We all too frequently judge a cat's social needs as being similar to our own, so many believe that keeping just one cat would be tantamount to cruelty. After all, wouldn't he be lonely?

In years gone by it was a common belief among companion-animal behaviourists that the cat 'walked alone'. It was considered to be an asocial creature and any attempt to create a group structure would be, at the very best, a tolerable but less than satisfactory situation for the individuals involved. As more and more research has been conducted into cat behaviour, mainly by studying groups of feral cats, it has been established that their social structure is somewhat more complicated.

A group of cats, referred to as a colony, consists of (at the very least) a queen and her kittens. These kittens remain with their mother and, if food resources are limited, they will move on once mature and the family will disperse. Larger colonies are composed of several females and their offspring, all of whom cooperate with one another in the rearing of their kittens by assisting at the birth, guarding kittens, helping with food provision and various other reciprocal activities that ensure the survival of the various litters. These larger colonies form around plentiful food sources such as food storage areas, the backs of hotels or hospitals or rubbish dumps that attract rodents. The adult males within the group will stay largely on the outskirts and may even have a home range that overlaps with other colonies. Some of the males tend to hang around just one colony and develop social bonds with the females. This is quite a successful strategy as females are more likely to mate with familiar males rather than those from outside the colony. Those males that remain with one colony will often show cooperation with each other when a queen is in season, rather than actively fighting, and also exhibit social behaviour among themselves at other times.

Other males will not associate with particular groups and tend to roam over greater distances and mate with females from a variety of colonies. It makes sense that the most successful of these males are large intimidating ones, because a small, weak one is unlikely to mate with a queen that has a number of familiar males around her.

Related cats tend to be the most sociable within a feral

colony, even more so than those that are merely 'familiar'. An element of feral colony life that is relevant to our domestic multi-cat groups is the strong resistance to the introduction of a strange cat, which brings about a great deal of social disruption if that cat chooses to stay.

So what happens when cats live with us in groups? We neuter them (usually) and we bring home the individuals we choose to group together; they may or may not be members of the same family. They then find themselves in a territory potentially containing other individuals, living in other properties, that would not necessarily have chosen to congregate in such close proximity. We then restrict, either partially or completely, their ability to explore and defend their territory outside or, worse still, restrict them to a 'territory' that consists of the rooms within the four walls of our homes.

One very user-friendly feature of the domestic feline is its incredible adaptability. Cat lovers will often collect an assortment of furry friends (as I did) and plonk them all in a house with little regard for the appropriate social niceties and, somehow, it works. Or, at least, it seems to work . . .

Some multi-cat households are great, with members showing genuine sociability among themselves. Unfortunately, though, cohabiting can instead be a matter of great stress for many cats, so the purpose of this chapter is to explore their social structure, how they communicate and how we can best play a positive role in ensuring everything goes well.

Territory

When we talk about our homes we are actually talking about a kind of territory. We see our house and garden (if we are fortunate enough to have one) as our castle and grounds, safe from invasion and worthy of defence should anyone try to come indoors and take over. We feel 'at home', relaxed and secure. Our home town, to us, is also a familiar place, but we don't challenge others for being there, because it doesn't belong solely to us. It's not without tension, though; how many times do you see a person approaching and avert your eyes to avoid direct communication? We venture outside our home town regularly, into the 'world beyond'; where we go may be familiar but ever changing and we must be on our guard and behave appropriately as it's certainly not home. Some of us are very nervous about going to a big city and feel distressed by the vast numbers of people invading our personal space and rushing in different directions.

Imagine what it would be like if humans were even more suspicious of strangers than they already are. Your life would be all about defending what is yours and attempting to avoid others, as their motives could well be malevolent. Imagine if you only felt safe in your bedroom because strangers could come into your house and use it as their own because they see it as just somewhere to hang out. So, when you step out of your bedroom, you never know whom you may bump into. You would go into your garden only rarely because it is full of dangerous people lurking in secret corners. How hard it would be to live like that! Welcome to the world of your cat.

Research into the lives and activities of feral colonies and domestic cat households shows us that a cat's world is divided into three recognized areas:

- the 'core' area or den
- the territory
- the home range

The core area

This is the area within which the cat feels most secure; sleeping deeply, playing, eating and enjoying social interaction. As owners represent the largest element of security, the core area is almost automatically the home itself. However, the size of the core area can be significantly smaller and may consist of a single room or even a high shelf. This could be for a number of reasons – for example:

- The cat is timid or anxious.
- There is a cat flap compromising the security of the entire home.
- The cat is in dispute with others in the household or in the territory.
- There is a lack of social stability in the household – e.g., people coming and going and exhibiting disruptive or noisy behaviour.

The territory

While the core area forms the hub of the territory there is an area beyond it that the cat actively defends against invasion by others. The size of this area varies greatly with each individual and depends on the season, level of confidence, sex, density of population, and many other factors. The boundaries of the owner's property do not automatically constitute a territory, as this will often include roads, wasteland, other people's gardens and even their houses.

The home range

This includes the territory but describes the total area over which the cat roams. In a study conducted by Roger Tabor between 1976 and 1978 in east London he found that the females involved had a home range of the garden plus any further space that their confidence and the density of population allowed. A typical home range in a densely populated area is 0.05 acres for a neutered female; neutered males often roam further.

Social communication

As cats hunt alone they actively avoid contact with other felines when they embark on hunting expeditions. Most of the cat's communication, therefore, is about increasing the distance between individuals rather than encouraging approaches. Those signals that are used to encourage interaction are reserved for familiar cats within the same social group.

Feline communication is divided into three groups:

- olfactory (using scent)
- visual (using body language and facial expression)
- auditory (using vocalization)

Olfactory communication

If you are attempting to increase distance between yourself and members of your own species, then it is best to find a method that doesn't rely on the recipient of the information getting 'up close and personal'. If you don't want your potential enemy to get near enough to see or hear you, you are left with little choice or alternative means to get your point across. What better method can there be than leaving a 'calling card' of your unique smell that is 'readable' by any passing felines long after you have left the area?

The cat is equipped with an additional organ of scent in the roof of the mouth called the vomeronasal organ (Jacobson's Organ), which enables the cat to 'taste' the smell and glean the maximum amount of information from it.

Cats have scent glands on their faces, flanks, tail base and paws and each area can be used to deposit scent as a signal to others. Scent is also useful for the depositor to communicate messages back to himself indicating the relative security and familiarity of particular areas. This marking behaviour is an essential part of the feline communication system and, by the very nature of its form, virtually unreadable to humans. It's important, though, that we try to understand a degree of the cat's motivation.

Rubbing Cats will use the glands on their faces, flanks and tail base to deposit scent on objects within their territory. You may observe your cat rubbing his face on furniture round the house or on your shoes, for example. He may also rub his cheeks against twigs in the garden. This is believed to relate to his sense of familiarity in a particular environment or social situation. Parts of the facial pheromones secreted by the glands in the cat's cheeks that are common to all domestic cats have been synthetically reproduced to treat a variety of behaviour problems.

Scratching Scratching is used for claw maintenance but also has a marking function, utilizing the scent glands between the pads. This enables both a visual (scratch mark) and olfactory message to be left in significant areas throughout the territory. These marks can be found on the edges of the territory as well as in important areas within it. If a cat scratches excessively in one area within the home, this may have strategic importance in some territorial dispute rather than purely being conscientious claw maintenance.

Urine-marking This form of marking is referred to as

'spraying'. The cat backs up towards a vertical surface and then, in a standing position, lifts its tail upright. A small jet of urine is ejected while the cat treads with its back paws and the tip of its tail twitches or shivers. Cats may even squat to deposit a small urine marker on horizontal surfaces. Research has shown that cats can distinguish between urine deposited vertically and horizontally, so it is possible that secretions from the anal glands are involved when the cat sprays; the urine certainly appears more viscous and oily than normal.

All cats – male, female, neutered and entire – are capable of spraying urine in the right circumstances. In sexually active males and females the urine mark represents an invitation that communicates their readiness for mating. In neutered cats it has the opposite purpose, as it deposits a scent that enables territory to be utilized by a number of cats without coming into direct contact with one another. The freshness or otherwise of the scent indicates when the other cat deposited it and therefore whether it is safe or not to enter the area.

Cats should have no need to spray urine indoors in a domestic setting, since this should represent a safe haven for the residents. If there is tension or conflict between the members of the group, though, they have limited ways of expressing this intense emotion and often resort to urine-spraying. The spray marks are focused on areas where the cat feels particularly vulnerable. Spraying can also be used as an attention-seeking action or in situations when the cat is frustrated. Passive aggression can also be associated with urine-spraying under certain circumstances, so clearly the practice is a complicated and even multi-purpose method of communication.

It is interesting to note that a survey into cat behavioural problems conducted in 2000 indicated that the incidence of urine-spraying indoors increased in direct proportion to the number of cats within the household to a level of 86 per cent likelihood in households containing seven or more cats. There is, however, a threshold for the number of cats in any group that, if exceeded, will inhibit the expression of such intense emotion and urine-spraying will cease. This is not a good sign.

Middening This refers to cats depositing faeces as a form of marking behaviour. It tends to take place at the boundaries of the territory or in open spaces of particular strategic importance, such as established pathways or 'cat runs'. If, for example, faeces are deposited in the middle of lawns or anywhere exposed, it is possible that this is a marking gesture rather than normal elimination. Middening as a single manifestation of olfactory marking in the domestic home is, in my experience, relatively rare. Urine-spraying and other obvious signs of tension almost always accompany it.

Visual communication

There will always be occasions when cats see one another, so their body postures have to be a clear signal of their intentions. Cats always attempt to avoid conflict, as approaching every encounter with 'guns a-blazing' would be a very poor strategy for survival. As a solitary hunter a cat could be severely debilitated by a fight injury and therefore unable to feed.

Cat A's response to the sight of another, for example, is to turn the body sideways and arch the back and raise the tail vertically. The hairs along the spine and on the tail become erect, giving the impression of a larger than average cat, and Cat B (the approaching cat) thinks better of it and moves on to other pastures. An alternative response, if Cat A is a less than confident individual, is to crouch down and create as small a silhouette as possible to fool Cat B into believing he, Cat A, is not there at all. Some cats, in response to these approaches, will stick their head under something nearby that can be used as camouflage. The poor souls often have large parts of their anatomy still in full view, their belief clearly being 'I cannot be seen if I cannot see'.

If the witch's-cat posture fails or the cloak of invisibility isn't quite so effective as usual, then Cat B will advance further. Avoidance of conflict is still Cat A's aim but running away at this point would undoubtedly provoke a chase. As there is no absolute guarantee which cat is faster it is necessary to employ further strategies. Cat A will start to walk very slowly away, almost in slow motion, with the body sideways on and the head slightly lowered and turned away, while remaining acutely aware of exactly where Cat B is. If Cat A is lucky he will move out of sight, find sanctuary and live to see another day.

Sometimes when we see two cats in close proximity to each other and clearly about to embark on something nasty, one of the cats will be on its side with its legs raised in the air in the direction of its adversary. This is often erroneously interpreted as a submissive posture, akin to that used in social signalling between dogs. As cats are not obligate social creatures, there are no innate postures for appeasement or submission and this particular behaviour is associated with defensive aggression in response to an imminent attack.

When a cat is challenged, its body positioning changes in subtle ways to indicate offensive aggression or fearful defensive aggression. The angles of the head, tail and ears all alter, sometimes dramatically, to signal specific emotions. The legs straighten or bend and the body flattens or arches depending on the cat's response to the challenge.

Paul Leyhausen originally created the following two charts

for his book *Cat Behaviour: the predatory and social behaviour of domestic and wild cats* (Garland STPM Press: New York, 1979) to show the varying degrees of offensive (aggressive) and defensive (fearful) posturing and facial expressions. Nothing has been produced to beat them so they are worth reproducing here. The bottom right-hand corner of the chart below depicts the ultimate in arousal and mixed messages, as the cat is signalling both offensive and defensive behaviour; this is probably a useful strategy to adopt when a fight is almost inevitable, to wrong-foot your opponent.

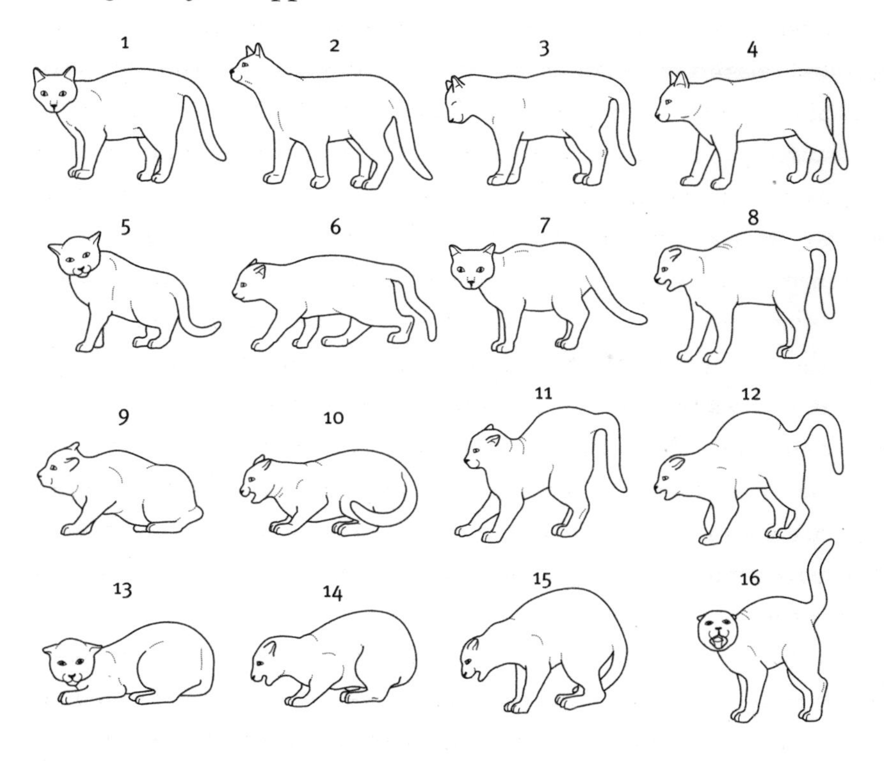

1 represents a calm cat
From 1 across to 4, the cat becomes more offensive
From 1 down to 13, the cat becomes more defensive
4 represents the most aggressive cat, being both offensive and assertive
16 represents a cat with defensive and offensive behaviour

Eyes Facial expressions are also excellent indicators of the cat's intentions. Pupil dilation is a sign that the sympathetic nervous system that fuels the fight/flight response has kicked in and the cat is prepared to face danger. When a cat has wide eyes and fully dilated pupils (it looks as if his entire eyes are black), it indicates that he is highly aroused and ready to defend himself if necessary. It doesn't necessarily mean he is frightened; it could just mean that the little toy mouse you are wiggling backwards and forwards in front of him is sending him into a frenzy of playfulness.

As a general rule narrow pupils mean that the cat is contented and relaxed, while dilated pupils mean that he is highly aroused. Staring eyes (fully open, unblinking and focused on a particular subject) are considered to be a significant challenge in cat-speak. Since human communication is all about direct eye contact, when we start a relationship with a cat a complete misunderstanding may result. On meeting a cat for the first time it is best to have relaxed eyes that blink slowly; this response is used between cats for reassurance in difficult situations. Many owners refer to this as the cat kiss, so feel free to display half-closed, slow-blinking eyes at any time when approaching a cat. Should you attempt to read your cat's eyes for a hint to his emotional state, don't forget that pupil size also varies according to the ambient light, so it must always be judged with this in mind.

Ears Ears, too, indicate different emotions. If a cat pulls them close to his head and folded downwards this means that he is trying to avoid confrontation. If the ears are flattened against his head so you can't even see the shape of them, this shows that he is probably going to attack because he feels he has no choice. If the ears are then flattened with a backwards rotation, watch out, because this cat is angry and about to strike. You will see the head draw in, the neck visibly stiffen and the mouth clamp shut. If this is directed at you and you are still there watching this with amazement, I would suggest you turn slowly and leave or face the consequences.

Tail We all recognize the friendly tail; it is held vertically and used as a greeting and, often, invitation to stroke. This tail posture is used between cats and often precedes mutual rubbing. The puffed-up tail with all the fur standing on end indicates a fearful or defensive aggression, as does the erect fur along the spine. A tail with an arch at the base but held downwards indicates offensive aggression. The tail is also often seen twitching and wagging from side to side. This has traditionally been interpreted as showing anger, but this is not strictly true. The wagging tail denotes a sense of conflict in the cat's emotional state that may well result in aggression but this isn't an inevitable consequence. Cats may also wag or twitch their tails when they are in pain or discomfort.

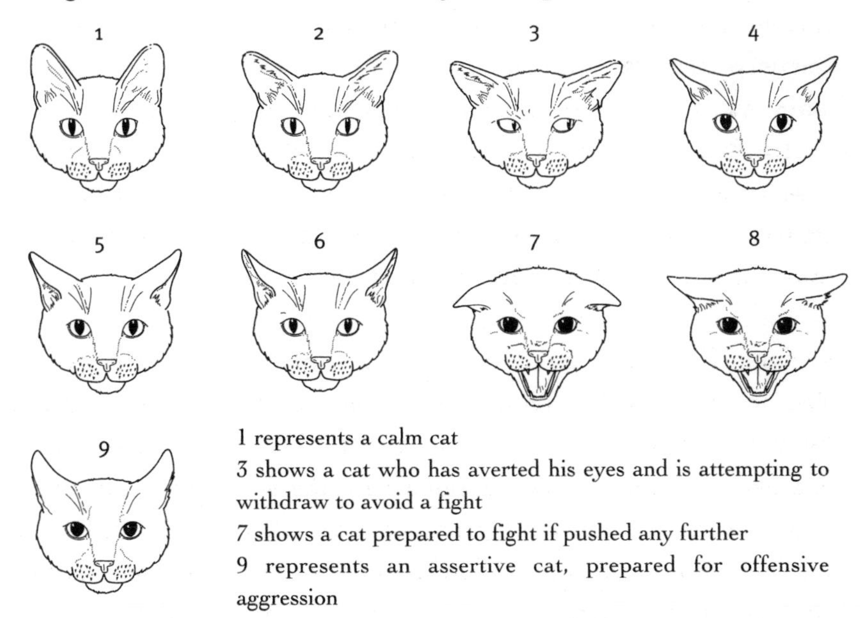

1 represents a calm cat
3 shows a cat who has averted his eyes and is attempting to withdraw to avoid a fight
7 shows a cat prepared to fight if pushed any further
9 represents an assertive cat, prepared for offensive aggression

Auditory communication

Most of what our cats 'say' to us is learned behaviour that they acquire over the years as a result of our response. Most old cats will have a full repertoire of noises that can be specifically interpreted as 'Open the door', 'Feed me' or 'Brush me' by loving owners everywhere.

Most natural vocal communication takes place between mothers and their kittens or is used in adulthood to greet others in a social context. There are several suggested ways to categorize vocalization either by differentiating between adult and kitten sounds or by getting slightly more technical and using sonographic analysis and categorizing them accordingly. They can also be grouped according to how the mouth is used to create the sound. This groups together those sounds we most associate with the 'miaow', used largely in a positive social context. These are created, much like human speech, by opening and closing the mouth. This category of vocal sound also includes the harsher 'round-vowel' call that occurs when a cat is prevented from doing something in particular or is in pain or even when it is about to regurgitate food.

The second group of sounds, including the purr and the chirrup, are made with the mouth shut. The purr is a greatly misunderstood sound, as many believe that it signals contentment. Unfortunately many sick and dying cats will purr and it is now understood that the purr signals receptiveness to social contact. (It could even be more complicated than that, as research conducted into the healing properties of low-frequency vibration shows that the cat's purr vibrates at a frequency that is considered therapeutic for bone growth, fracture healing and pain relief.)

The third group, with the mouth fixed open, comprises those sounds used in conflict, including the growl, hiss, shriek and spit. One of my old cats, Bln, now gone, was a great exponent of the shriek during encounters with other cats in the neighbourhood. In his endeavour to avoid conflict at all costs he would, at the point of contact, shriek like a banshee; then, when his opponent immediately pulled back in amazement, Bln would take the opportunity to escape.

Social maturity

When we view kittens playing together and cuddling up to sleep cradled in each other's paws it is hard to believe that so much cat communication is about conflict and conflict avoidance. Many owners are fooled into thinking that all cats are sociable with one another because they have seen this in their own kittens. However, if they stopped to think, surely they would be a little suspicious as most kitten games are about fighting and antagonistic posturing. The reality is that these kittens grow up and, in the part of their lives outside that you don't see, spend hours avoiding conflict and danger. Even when they come back indoors they are not quite so lovey-dovey as they used to be with one another. The owners just don't see anything because it's been a gradual change and there is no fighting.

Social maturity occurs between the ages of eighteen months and four years. Some breeds, such as the Siamese, can be quite precocious in their development and acquire a strong sense of social awareness after one year. Most cats mature after two years of age (beware the Burmese's second birthday!) and it is then that you will see the cat's true response to its own group and other cats in general.

At this time the relationship between the members of a multi-cat group may be pivotal to the response of the cat that has recently matured. If there is a familial connection, then this may well influence the cat's response to the others. There may be trouble if the rest of the cats coexist without any genuine bonding – i.e., they share the same territory but not much else. Even if you haven't noticed anything particularly untoward, your cats will have adopted one of the following strategies.

Agree to disagree, share the same space but avoid direct contact

This will potentially be the tactic employed in households where the owners have noticed no obvious change to the relationship. Cats will continue to share the house but the space will have been divided into various core areas, passages and thoroughfares with some specific locations – e.g., feeding station, cat flap, litter trays – being shared. These will be the places where friction could occur depending on how motivated each cat is to protect these important resources. For example, one cat may love its food and be pushy and aggressive at meal-times, whereas the other may stand back and let him get on with it. This doesn't necessarily mean he has been scared into submission; it may just mean that he isn't that excited about food.

The cats that have adopted this strategy are to be found in separate parts of the house when resting. They have very definite preferred sleeping locations and other cats will never be seen there (unless there is a problem or a particularly good relationship going on). The exception to this rule are the finer-coated pedigrees – for example, the Siamese or Burmese – who are almost invariably to be found huddled up to a source of heat even if it is another cat. Unfortunately, this doesn't necessarily mean that there isn't a strong sense of competition in other situations.

These kinds are probably worthy of the phrase 'good social

groups'; I would leave well alone and resist the temptation to add more cats. There is no guarantee they won't fail to integrate and even cause friction between the resident members.

Be oblivious to the presence of other cats

There is a certain personality type ideally suited to cohabiting with others. Some owners describe these cats as 'a bit thick', 'laid-back' or 'chilled'. They are certainly not remotely stupid but of a type that is sufficiently content to exist in any social situation without letting underlying tension get to them. I have seen many households where there have been real issues (fighting, urine-spraying, entrance-blocking) between certain members yet one cat walks around apparently oblivious to it all. This cat will occasionally get caught in the crossfire and receive a wallop or a spit in the face. The response is one of mild surprise, a little face pulling, and then off he goes to find a warm spot for yet another bout of relaxation. I think these cats should be cloned for all those people who feel they could never have too many and insist on filling their homes with furries.

Continue to have a truly sociable relationship

Despite the bad press I so often give them, multi-cat households can work. If there are no troublemakers outside, then it is entirely possible that social maturity will bring no adverse effect. Cats will continue to greet each other with genuine affection and all will be well. The highly social Oriental breeds often develop real attachments to other cats and this frequently continues for life. There is some debate whether or not these relationships are, in reality, co-dependencies, with both parties needing rather than wanting social contact. You might say, 'What's the problem? They seem friendly enough.' It isn't necessarily a problem until one of them dies, gets ill, goes to the vet or in some way isn't there for the other one. This can cause a serious 'cold turkey' effect as the remaining cat struggles to cope.

The above are the three main positive results of social maturity. If all your cats are mature, and if the occasion has come and gone with no hiccups, you can safely assume it's going to be OK. Beware of making that assumption too soon, however; all members need to get there before you can announce that you have the perfect set-up.

Fail to cohabit successfully

Unfortunately, given the job I do, I have more contact with this type of group than any other. There are a number of strategies that a cat may employ under these circumstances:

- The cat despises the others and embarks on a battle of psychological threat and intimidation to destroy the enemy.
- The cat still despises the others but is too scared to show it (and probably eventually develops a stress-related illness).
- The cat is terrified of the others and gets beaten up with monotonous regularity.
- The cat votes with its paws and moves out.

Social hierarchy

For many years now researchers and behaviourists have tried to get a handle on feline hierarchies. Are there such things and, if so, what form do they take? Many theories existed – for example, that the cat has a linear hierarchy headed by an identifiable 'alpha' cat. When this cat dies the next cat in the 'chain of command' takes over and adopts the role. Another theory holds that there is an 'alpha' cat and the remainder of the group represent 'middle rankers', all of whom defer to the General at the top. Unfortunately, a great deal of the research carried out had many variables that were not taken into consideration when conclusions were drawn. More recent thinking suggests that hierarchy doesn't exist at all. Social groups arise as a result of cooperative behaviour and the harmony is maintained via mutual rubbing and grooming to uphold the communal scent. It is certainly true that you will see a degree of cooperative behaviour in households as one cat sits back while another eats or one leaves an owner's lap as another approaches. But is this truly cooperative behaviour or an indication of one cat deferring to another in a particular situation?

I have studied many domestic cat groups, some perceived as good and some bad. I would suggest that often you can identify what some consider to be a 'top cat'. This is the one that distances itself from the group and rarely interacts with others. This cat will sit where he likes, use the litter tray as often as he likes and have access to you, even if others have to leave your lap to make this possible. These cats tend to be confident individuals and it would be more accurate to describe them as assertive rather than dominant. There is a big movement away from using the adjective 'dominant', even when referring to dogs, as the whole concept of the dog as a pack animal is considered now to be far too simplistic.

There are certainly a few elements of multi-cat living I have observed over the years that appear to hold true for every group I have seen:

- Anxious cats always defer to more confident cats in competitive situations.
- The outcome of one antagonistic encounter may shift the

'balance of power' between two equally matched cats.

- Not all cats are as territorially sensitive as others.
- A multi-cat household may actually consist of a number of 'splinter groups' that just happen to inhabit the same territory.
- Not all cats that cohabit with members of their family can comfortably live together once they mature.
- Cats can be bullies: if a member of the group behaves like a victim it will be treated as such and picked on mercilessly.
- The response to an act of aggression dictates whether or not the behaviour is repeated – for example, if a cat picks on another and gets a good walloping for it he probably won't try that again!
- Some relationships between cats that look very loving are probably co-dependencies between needy individuals.
- A degree of active aggression (hissing, etc.) is inevitable in a multi-cat group under certain circumstances and should really be considered normal.
- Direct staring is challenging and usually employed by more assertive individuals.
- Highly confident cats often refuse to fight actively. Instead they look away and walk away, then sit and groom, indicating that the other cat has lost.
- Mounting behaviour can be employed as an assertive territorial gesture.
- The hiss is used to avoid active fighting.

Signs of unrest

A group of cats that are mutually tolerant can be considered as acceptable, while a group at war are clearly not. It's best to know if there is chronic tension in your household, but recognizing it isn't as simple as watching out for fisticuffs, as cats 'at war' are not necessarily fighting with teeth and claws at every opportunity. They rely on avoidance strategies or, if necessary, passive and covert tactics designed to disarm the enemy emotionally.

The following points are not necessarily diagnostic of a major problem in need of intervention but they could give you a better insight into what is really going on. The medical issues and the recognizable 'behavioural problems' such as urine-spraying are certainly the ones that should sound the alarm bells.

- Individuals keep away from one another within the house and do not sleep or rest together.
- Members of the group respond to the presence of another by leaving the room.
- The cats will not play in the presence of another.
- One member appears to withdraw from human contact in the presence of another.
- One cat will move another away from a favourite resting place just by staring.
- One member of the group will sit in passageways, by the cat flap or on the staircase, preventing other cats from moving freely around the house.*
- There is active fighting in narrow corridors or at mealtimes.
- There is a fight when one of the cats returns from a visit to the vet.

* This can be particularly telling. If you want to try a little diagnosing, draw a ground plan of your home with windows, doors, staircases, cat flaps and all favourite cat things (litter trays, etc.) marked clearly. Then study your cats for a while and make a note of where each one prefers to spend most of his time. If a particular resting area hinders movement round the home or blocks access to the cat goodies, it may not be a random choice of location. Watch the response of the others; it might be very revealing.

- There is evidence of excessive scratching in certain areas in the house.
- At least one member of the group is obese.
- There is a history of inappropriate urination or defecation (soiling) in the home.
- There is a history of urine-spraying indoors.
- One member of the group over-grooms and gets bald patches.
- One member of the group suffers from recurrent cystitis.
- One member of the group spends prolonged periods away from home or has to be retrieved from houses several streets away.

Positive social signs

There are some strong indicators of friendliness and sociability between cats, so, as reassurance to those who are now slightly worried, it's important to recognize these too:

- chirrups when greeting one another
- sleeping together
- grooming each other (referred to as 'allogrooming')
- rubbing against each other to exchange scent (referred to as 'allorubbing')
- playing together

There are certain things that you can do to limit the chances of your multi-cat household going badly wrong. You can't influence, for example, the number of cats that suddenly appear in your garden but you can ask the right questions about the cat population when you move to a new area. You will never remove the risk completely but you can limit the odds.

- Choose compatible individuals such as littermates,

probably brother and sister, when you first acquire cats. Two males of equal age may fight when they mature socially.

- If you want to adopt an adult cat to add to your group, choose an individual that shows a history of being sociable with other cats. Avoid those that have been given up for adoption because of indoor soiling/spraying/anxiety-related problems. Many cats enter rescue centres with very little information on their background; acquiring a second-hand adult cat always represents a degree of risk. If you have a stable group already, why risk it?
- Avoid extreme characters – e.g., those that are extremely nervous or confident – when choosing new kittens or adding to an existing group. These may potentially be difficult cats to live with or be those that find others difficult to live with.
- Don't be under the impression that keeping kittens from your own cat's litter will be company for her. Once the weaning process is complete there is no guarantee that they will cohabit successfully.
- Keep the appropriate number of cats to suit your environment – for example, if you have a two-bedroom cottage you may be overdoing it with five. This is particularly relevant if your cats are kept exclusively indoors as there is little opportunity to avoid one another should they so wish.
- Try not to constantly add to a stable social group. In my experience, every household has a 'one cat too many' threshold and the introduction of one disruptive newcomer will upset all of them to some extent.
- Avoid too many cats in your household in densely cat-populated areas. A sense of overcrowding can induce stress even in those cats kept as singletons. The relationship between members of a group can be environment and

territory specific, so think ahead when planning to move house.

- Avoid feeding stray cats and encouraging them into your home. They may seem gentle and vulnerable but they can have a very strong sense of territory and attempt to turf your cats out once their paws are firmly under the table.
- Avoid too many highly intelligent and sensitive pedigrees in the same household. They can be extremely territorial and competitive with one another, particularly breeds such as Burmese, Bengal and Siamese.
- Provide dry food for 'grazing' throughout the day or to avoid any sense of competition divide it into several smaller meals available only at certain times. To avoid bullying at mealtimes designate several areas within the home as feeding stations. (This method of feeding is not recommended for those cats suffering from feline lower urinary tract disease or chronic cystitis; they need to be fed a wet diet.)
- Water is an important resource to cats and several bowls throughout the home will potentially encourage them to drink more frequently. Cats find water more attractive if it is located away from food. It is advisable to encourage your cats to drink as much as possible to guard against urinary tract problems.
- Provide plenty of opportunities for your cats to perch in high resting places. Cats often prefer to observe activity from an area they perceive to be safe.
- Private places are also extremely important; every cat, no matter how sociable, needs 'time out' to enjoy moments of solitude. Wardrobes and spaces under beds are ideal and there should be plenty of choice to enable all members of the group to have their own favourite place.

- Ensure there are plenty of scratching posts to prevent damage to your furniture. Cats will scratch for both claw maintenance and territorial reasons and there will be an increased need to signal to others in a multi-cat environment. These scratching posts should be located near entrances, exits, beds and feeding stations to ensure an appropriate surface is available in areas of potential competition.
- Beds in warm places are worth defending, so provide enough for everyone.
- Even if your cats have access to outdoors it is still advisable to provide indoor litter facilities. If there is any bullying going on outside, then your cats will always have the choice to toilet in comparative safety indoors. Providing 'one tray per cat plus one' in different locations indoors is an important formula if you already have a soiling problem in your home or a cat with a history of urinary tract disease. However, prevention is better than cure and ensuring the availability of a number of convenient litter trays will potentially avoid problems in the future.
- If you feel there is any tension between your cats, it would be advisable to provide two separate entry and exit points to your home – i.e., cat flaps, doors or windows – to avoid the risk of guarding or blocking and to enable even the most timid cat to get indoors unhindered.
- Food, water, litter trays, beds, toys, scratching posts, high perches and private places are all really important to cats. If you want to limit competition between your cats over these precious resources, then provide them according to the formula 'one per cat plus one extra positioned in different locations'.

CATS, BABIES AND PREGNANCY

It isn't always other cats that enter your home. The introduction of any new family member will have an impact on your cat, so preparation is the key. There is a great deal of myth regarding the dangers of mixing cats with babies. You will still hear people saying that smother cats babies by sleeping on their faces or they carry dreadful diseases that could harm the unborn child. This can cause a great deal of unease in mums-to-be and many wonder whether it would be safer to re-home the cat for the wellbeing of the baby. I'm a great believer in finding out as much information as possible about things that scare you. It is often the lack of knowledge that fuels the fear, so having the facts will enable you to make the right decisions when the time comes.

Pregnancy and toxoplasmosis

One thing that you really need to understand is the actual threat posed by one particular parasite. There has been a great deal of publicity about the risks of contracting toxoplasmosis during pregnancy. Toxoplasmosis is a disease caused by infection with a parasite called *Toxoplasma gondii* (T gondii) found in raw meat, cats that eat raw meat, and cat faeces. If a woman is infected with T gondii during pregnancy there is a risk that it will affect the unborn baby, causing stillbirth, miscarriage or congenital defects.

Approximately 20 to 30 per cent of people in the UK have been infected with toxoplasmosis at some stage in their lives. In people with healthy immune systems the symptoms are mild (and often go unnoticed) or cause fever and swollen lymph nodes in the neck. Those who have been infected with T gondii develop antibodies to the organism and there is no risk that the infection can be passed to the foetus during pregnancy. However, it is estimated that 1 in 500 women in the UK contract toxoplasmosis in pregnancy, with only 30 to 40 per cent of those passing the infection to the unborn baby. The risk varies depending on what stage of pregnancy the woman had reached when the infection was acquired.

Sources of infection

In most cases, people become infected via one of two routes:

- ingestion of oocysts (part of the life cycle of the parasite) from the environment, either directly through contact with contaminated soil or indirectly by eating contaminated vegetables or fruit
- ingestion of fresh meat containing tissue cysts

Cats have had a pretty bad press regarding this issue but recent research indicates that contact with cats does not increase the risk of T gondii infection in people. According to the studies it is rare to identify cats shedding oocysts in their faeces.

Precautions It certainly isn't worth taking any risks, however, so a few precautions should be taken:

- Avoid cleaning out litter trays during pregnancy.
- Avoid gardening or handling of soil.
- Only eat meat that has been cooked thoroughly.
- Wash hands, cooking utensils and work surfaces after preparing raw meat.
- Wash all fruit and vegetables thoroughly before eating.

Preparing your cat for the arrival of your baby

- Get information from your doctor or health visitor about any risks to your unborn baby from organisms known to be carried by cats and take any precautions advised by them.
- Get up to date with vaccinations, worming and flea control for your cat. Flea-bites are not life-threatening but they are incredibly unpleasant for a baby, so you want to make sure there are no fleas around when he or she arrives.
- If you haven't got round to neutering your cat, do it now without delay.
- Now is the time to address any existing behavioural problems that your cat may have, even if you have tolerated them before. The upheaval and disruption caused by the arrival of a baby may make the problems worse.
- Make the decision about where the baby is going to sleep at night. If the cot is initially going in your bedroom, then start denying access to your cat early in the pregnancy. Spray synthetic pheromones ('Feliway', available from your vet) against the door frame if your cat reacts to the closed door and seems distressed.
- Ensure that the designated nursery is also made out of bounds at this stage so that it doesn't represent another change once the baby has been home for a few months.
- Introduce all the baby accessories such as buggies and cots over a period of time to avoid a sudden burst of challenging smells and objects. The pheromone spray Feliway can also be applied to these items to minimize their impact. You may also want to purchase a cot and/or pram net. This will deter your cat and any flying insects.
- Bear in mind that the baby's arrival will have a greater impact on your cat if he is kept exclusively indoors.

House-bound cats are more sensitive to changes in their environment.

- If your cat is used to having your undivided attention it's important to withdraw gradually from him during the pregnancy. No matter how much you love your cat it will be virtually impossible to sustain the same degree of attention once the new baby arrives. Start to plan the gradual withdrawal from your cat as soon as your pregnancy is confirmed. Provide more stimulation for him – for example, more time outside or more activity indoors that will give him an outlet and interest beyond his relationship with you. Imagine what your new responsibilities will be with the baby and create a timetable and new cat routine that will slot in nicely, including feeding, grooming and playing. If this routine is adopted as soon as possible it will reduce the impact of baby's arrival.
- Encourage friends with babies to visit so that your cat gets used to their presence. Take care if small children are also present at these times as cats can find them overwhelmingly noisy and persistent. Always supervise encounters and ensure that any handling is gentle and appropriate. It's important to make the experience pleasant by using the positive reinforcement of food treats when children are around. If, however, your cat chooses to hide at these times, don't worry and don't pursue him or force him to interact.
- Do not comfort him if he appears frightened of children; this is an inappropriate fear that should not be reinforced.
- Plan ahead and ensure there are plenty of high resting places where your cat can retreat away from baby when he or she starts to crawl.

Your baby's arrival

- Try to stick to the cat routines you established during your pregnancy.
- If you find you can't cope, ask friends and family who know and love your cat to pop round and play with him. Your partner can help by paying greater attention to the cat while you are busy with the baby.
- Create private spaces and secret nooks and crannies for your cat, just in case he finds everything overwhelming and wants to escape when baby comes. This is perfectly normal and it's best to respect this and not bother him while he's in retreat. He'll soon come round! Wardrobes, cupboards, cardboard boxes with soft beds, all are useful for this purpose.
- Your cat may or may not be curious about the new arrival. If he wants to sniff the baby, don't panic! Introduce them in a quiet room where the cat has few associations – not in a place where the cat usually sleeps or eats. Reward your cat's calm behaviour with gentle praise and tasty titbits. If your cat prefers to run away from the baby, don't force him to come back; he will investigate in his own time.
- Keep the nursery or bedroom door closed when the baby is sleeping in his or her cot and, just before closing the door, make sure the cat isn't hiding somewhere in the room.
- Keep all the baby's feeding utensils out of reach of your cat.
- Try to keep calm; when you get overtired and anxious, both baby and cat will react by becoming tense themselves.
- Don't leave a new-born baby alone with your cat, no matter how trustworthy he is.

Introducing a new kitten

You may be thinking about expanding your four-legged family rather than your two-legged one. I hope you have thought long and hard about the implications of another mouth to feed and the ramifications of a new arrival for your existing brood. If so, and you still want to go ahead, you are now in the position to look practically at how you are going to make this work from day one.

A little bit of extra effort at the beginning can make the difference between a good or bad relationship in the future. Your existing cat (or cats) will have established his territory and the introduction of another, albeit a little kitten, is not necessarily going to be well received. It's important to ensure that the resident cats are not given the impression that they are under siege. Here are a few suggestions:

- Choose your new kitten with care, bearing your cat's personality in mind. For example, don't purchase a very confident and outgoing kitten if your existing cat is timid or shy.
- Arrange to collect your kitten on a day when you know you will have plenty of time to devote to settling him in – for example, Friday evening before a weekend off work.
- Before the kitten arrives, buy or hire a kitten pen and position it in a room that your existing cat doesn't particularly favour – for example, a spare bedroom. A kitten pen is a large metal cage with a solid floor and it is normally used for kittening queens, or post-surgery cats that need to be confined. It is quite large, with plenty of room for a bed, toys, food, water and a litter tray, and is easily collapsible to enable it to be moved from room to room.

- Close the door to this room and allow the kitten to exercise within it when your other cat is not around. The kitten's food, water, toys and bed can be positioned outside the pen but the litter tray should remain within it.
- The initial contact between kitten and cat should involve food. Open the door to the room while the kitten is eating food in his pen (with the pen door shut). Place a small bowl of your cat's favourite food as near to the pen as he will come to eat.
- Allow your cat to explore the pen without intervention and praise him if he starts to eat the food provided.
- Reduce the distance between the pen and your cat's food bowl by small amounts every time they meet in this way.
- Exchange bedding between the two to allow them to become familiar with the other's scent.
- Provide attention to the existing cat but do not exceed the amount that he finds enjoyable. Try to maintain his normal routines as much as possible to indicate that the kitten represents no loss of resources or enjoyment.
- Start to place the kitten pen in other rooms that are of increasing importance to your cat so that he will understand that the kitten has rights of access to all areas.
- Allow several weeks before opening the pen and letting the two get to know each other if you are unsure of your cat's response.
- If he appears gentle and curious, you may wish to reduce this period of time accordingly.
- When you do leave the pen door open or allow the kitten to leave its designated room, keep a cushion or pillow handy to place between them just in case things don't go according to plan. This is recommended only if you think that the kitten will actually come to harm; don't

intervene if what happens is just noise and posturing.

- As soon as the kitten is given access to the room where the litter tray is permanently located it's important he gets used to the idea of it being there. Consider shutting the kitten in this room at night to ensure he gets the message that this is the only place to go to the toilet.
- It may be wise to separate the kitten from the adult cat at night until the little one is physically more robust – i.e., approximately five months old – just in case your cat gets a bit too boisterous.

Introducing a new adult cat

You may be considering an adult cat instead of a kitten. This is always going to be potentially more difficult, but occasionally circumstances conspire against us and we feel compelled to accept a mature cat into our midst. Confinement in a kitten pen can be quite distressing for an adult cat, so this isn't necessarily the best method of introduction. I would recommend that the new cat is kept in a single room first rather than a cage; the resident cat can then be introduced gradually to the newcomer by following three basic steps. This single room should, once again, be an area where your resident cat doesn't spend much time.

Step 1 **Scent** Your cat should first be aware of the scent of a new cat. Cats have glands around their head that secrete a pheromone that signals a positive message of security and familiarity. The new cat's scent can be collected and deposited in areas where the existing cat is housed and vice versa. Scent collection is fairly straightforward using fine cotton gloves or a small natural-fibre cloth. Stroke the cat around the cheeks, chin and forehead using the cloth or glove. This will collect small amounts of the naturally secreted pheromones from the glands in the cat's face. The cloth can then be rubbed against doorways and furniture to enable both cats to explore the scent of the other without direct contact.

Step 2 **Sight** Your cat should then be allowed to see the new cat before he is able to make physical contact. A wood-and-wire frame to fit within the door surround can be useful for this purpose. Feeding either side of this barrier, at a distance that allows both cats to eat comfortably, is a good ice-breaker. Initially this distance may be halfway down the stairs as far as your cat is concerned, but moving the bowl nearer by inches at a time will help him to adjust and become less reactive to the newcomer's presence.

Step 3 **Touch** Physical contact can then be established after a reasonable period of time. What length of time turns out to be 'reasonable' will depend on the cats' individual characters. When they do first meet it is often tempting to interfere in their initial interaction, but unless they risk injuring each other it is usually best to let them sort it out in their own language.

The most difficult introduction to make is Number 2 cat to Number 1. The resident cat will have previously lived alone and is unlikely to embrace the concept of sharing. Introducing cat number three, four or five is usually less traumatic, but careful selection is still important. Avoid collecting cats from different familial groups that are similar ages and the same sex. Acquiring a good mix of male and female of a wide age range will prevent the problem of a number of cats all maturing at the same time.

Introducing a new dog

I have often said that some cat and dog relationships are great, despite what all the cartoons tell you. Introducing an adult dog to an adult cat may be difficult since many dogs, confronted with a disappearing cat, will automatically give chase even if they have no intention of doing harm should they catch it. If your cat has no experience of dogs this can be a distressing experience and many, given the opportunity, will leave home for a period of time before coming to terms with this drastic change in the household. The breed of dog chosen will also influence the future canine/feline relationship in your home. Terriers, greyhounds and other breeds designed to chase small furry objects are ideally to be avoided, as this combination is probably just asking for trouble. The breeds that are traditionally considered good with children such as the golden Labrador, retriever or Cavalier King Charles spaniel are probably sensible choices for multi-species living.

Introducing a puppy or adult dog to an adult cat

A puppy is easier to work with as it is young and malleable and will soon become used to the presence of another species, treating it as just another member of the family. However, it is probably still best to adhere to the recommendation above of 'suitable' breeds.

- Ensure there are plenty of high resting places in the home where your cat can retreat, away from the new arrival.

- Consider placing a baby gate at the bottom of your stairs to give your cat the sanctuary of the first floor.
- Introduce your puppy or dog to his new home by using a puppy pen or crate.
- If you are introducing an adult dog, then the pen can be used for quiet times and sleeping.
- Plan ahead and start to feed your cat in an area away from the location where you intend to put the puppy pen or dog crate; this will prevent your cat going off his food.
- Try to place the pen well away from the thoroughfare leading to the cat flap or normal exit route for your cat.
- If litter trays are provided indoors, ensure they are located discreetly and in areas where your new puppy or dog will not be able to go.
- Introduce the new puppy to your cat in a room where the cat can easily escape.
- Hold your puppy and allow your cat to approach, if willing.
- Your cat may hiss or growl but if you are holding the puppy you can protect him from any aggressive advances.
- Allow the cat to be in the room where the puppy's pen is located.
- When your puppy is out of the pen it would be advisable to keep a long lead on his collar to stop him from chasing your cat.
- Do not allow any unsupervised encounters until both parties are relaxed in the other's presence and the puppy has been trained not to chase.
- If you are introducing an adult dog, then place him in his pen with a tasty chew and bring your cat into the room.
- Give your cat attention at this time: grooming, play or his favourite food treat will help to create positive associations with his new canine pal.

- When your cat seems relaxed in the same room, you can open the pen (with your dog on a lead) and allow your dog to sit beside you while you hold the lead to prevent him from chasing.
- Reward your dog's calm behaviour with a food treat.
- Continue to have the dog and cat in the same room (with the dog on a lead) and give food treats to reward calm, relaxed behaviour.
- After a few weeks, depending on progress you can try without the lead.

Introducing a kitten to a dog

You may be the proud owner of a dog and decide you also want to share your home with a cat (wise choice!). If you do introduce a kitten you have to be aware that there is a major size difference between him and your dog, so unsupervised contact is out of the question until your kitten has grown enough to escape safely any unwanted advances.

- Initial introduction is safest with the kitten inside a pen.
- The choice of room is not as important as it is for a cat/kitten introduction, but the place should certainly be away from your dog's feeding area or anywhere he actively defends.
- Allow the dog to explore the kitten and vice versa with the latter in the pen to avoid the risk of injury or a chase ensuing if the kitten runs.
- The pen can be moved into each room that the dog has access to.
- If your dog isn't allowed upstairs it may be worth creating a cat sanctuary there so that your kitten will always know it can go somewhere safe.

- Position your indoor litter facilities and cat food with care; your dog will find both these things irresistible, so it's important he is denied access from day one.
- This process should continue for several weeks, particularly if your dog is protective about food, for example, and may respond aggressively when the kitten approaches his bowl.
- Once both seem relaxed in each other's presence, the kitten can be held near the dog with the dog on a lead to prevent chasing.
- Ensure the kitten is able to get away if he feels threatened.
- Treats can be given to the dog if he doesn't attempt to chase.
- It's important also to train your dog (if he isn't already) not to chase cats and to respond to your command at all times, just so you maintain control.

There are great sources of advice available on the Internet or through your veterinary practice. It's wise to follow the guidelines of those organizations that are considered to be bona fide authorities on their subject – for example, the Feline Advisory Bureau, the Association of Pet Behaviour Counsellors, Cats Protection, the RSPCA, or any of the other large pet rehoming charities. It's always best to be prepared with all the right information before you introduce a new family member. Cats are sensitive creatures socially, so it's essential to get it right from the start if you want a harmonious household.

CHAPTER 8

Feline Felons

A BRIEF A–Z OF COMMON BEHAVIOURAL PROBLEMS

I'M NOT GOING TO REPEAT EVERYTHING I HAVE ALREADY discussed thoroughly in *Cat Confidential, Cat Detective* and *Cat Counsellor*. If you have a problem with your cat's behaviour, there is no substitute for picking up all three books and reading them from beginning to end!

This is just a brief summary of the most common behavioural problems, with a few hints and tips indicating those things you really should do or consider and those you should definitely not attempt. All of these dos and don'ts are tried and tested and could save you valuable time by indicating the most effective course of action early on. The longer you leave a problem (we all hope they are going to go away), the more difficult it will be to resolve, but the original motivating factor for the behaviour is very easy to detect and simple to put right. Happily, too, some 'inappropriate behaviour' disappears without intervention.

Aggression to humans

This is a category of problem behaviour with numerous causes and motivations; it is really misleading and inaccurate to presume that if a cat displays aggressive behaviour to a human it is clearly an aggressive cat, end of story.

Just some of the motivations for aggression include:

- fear-related aggression
- play aggression
- misdirected predatory behaviour
- learned aggression (commonly used to deter unwanted attention from humans)
- frustration-related aggression
- redirected aggression
- maternal aggression
- idiopathic aggression
- aggression related to illness

All of these require tackling in very different ways. You will find if you ever try to obtain advice from a pet-behaviour counsellor over the telephone regarding an aggression problem that you will not be successful. All that any behaviourist worth his or her salt will do is give you basic advice to keep yourself and your family safe awaiting advice from your vet and formal behavioural intervention in the home (if appropriate).

Here are a few things that you need to know that could save you from further injury without the risk of making matters worse.

Don't bother with:

X Screaming and thrashing your arms about

This is easier said than done. A cat bite feels like a hard punch to the targeted area, followed by a transient numbness followed by excruciating pain. It is a natural instinct to withdraw rapidly from assault, but this can result in the cat biting again. Cats tend to bite, puncturing the skin with their long canines, and then withdraw immediately, so under these circumstances your showing the minimum of fuss is the best way to stop the cat coming back for more. The most important strategy immediately after the bite is to prevent further attack (shut the cat in a room and leave it for at least a couple of hours) and carry out first aid on the resulting wound.

A cat's mouth contains a disgusting cocktail of bacteria, including pasteurella, staphylococcus, streptococcus and clostridium, to name but a few. Small puncture wounds tend to heal quickly, trapping these bacteria under the skin. Some of them are anaerobic, meaning they don't need oxygen to multiply, so it is entirely likely that you will end up with an infection, possibly even septicaemia, as the bacteria spread throughout the bloodstream. Scary stuff.

Two things need to be done after you have been bitten: wash the wound thoroughly, ideally attempting to flush the wound out before it closes, then go straight to your doctor, who will prescribe antibiotics. Signs to look for include redness, swelling, heat, pain and possibly flu-like symptoms.

Cat-scratch fever (also known as cat-scratch disease) is also worth a mention here. It is an infectious disease occurring as the result of a scratch or bite from a cat carrying the bacterium *Bartonella henselae*. It causes swollen lymph nodes in children and occasionally fever, headaches, tiredness and loss of appetite in adults. It can be more serious in those people with compromised immune systems. It is usually self-limiting within three weeks, but severe symptoms can be treated with antibiotics. The incidence of this disease is reported as 6.6 cases per 100,000 population in the USA, so although it's unlikely to afflict you or your family tomorrow it's always advisable to clean any scratches or bites thoroughly.

X Punishment

Whatever has motivated a cat to show aggression, punishment won't improve matters. It will be seen by the cat as a further assault (if your initial approach was interpreted as threatening, resulting in the launch of a counter-attack) and you will end up being bitten again. It's important to remember, although impossible to rationalize this at the time, that not all aggressive attacks are about you. Some aggression can be redirected on to innocent bystanders when the cat is adrenaline-pumped having seen a strange cat in the garden, for example. Punishment will merely make an enemy out of you as well.

X Giving as good as you get

There is a school of thought that you have to show a pet who's boss 'because, let's face it, a cat that dares to bite or scratch its owner is getting ideas above its station'. Followers of this strategy are the same people who discipline dogs by grabbing them by the scruff, choking them with check chains and beating them into obedience and compliance. In other words: ignorant and cruel. If you decide to retaliate with a greater level of violence to prove dominance, you will not win and you are definitely not the sort of person who would read a book about how to care for your cat. Therefore, this piece of advice isn't for you but you can pass it on if you ever hear of anyone considering this strategy.

X Re-homing

Some aggression is situation and environment specific. Take the cat away from the environment or the relationship it is finding so difficult and it may never again have to express such a level of fear, arousal or frustration. However, there has to be a caveat here. It is irresponsible to re-home a cat with a history of aggression without either first exploring the motivation for the attacks with a professional or giving the re-homing centre full details of the whys and wherefores of all attacks or aggression shown. Cats with a history of aggression are not automatically euthanased as many people fear. These days the large cat-rescue charities have a team of behavioural experts who are experienced at assessing cats and able to re-home to the appropriate environment. If, however, the cat shows extreme aggression and the triggers are random or difficult to predict, it is virtually impossible to find a home where people will not be at risk of serious injury.

Try instead:

√ A visit to the vet to rule out a medical or pain-related cause

Cats do not suddenly show aggression for no reason. There are a number of medical conditions that can manifest themselves in aggression, including hyperthyroidism, urinary tract disease, arthritis, skin disease and even seizure activity and brain tumours. The aggression can be defensive when pain is present as the cat attempts to guard the damaged area.

It is often not the easiest thing to get an aggressive cat into his carrier for the trip to the veterinary surgery. Unfortunately, if you are now scared of your cat you will probably struggle; I would recommend stout gloves and a thick towel, or else ask the man next door to do it. Probably don't mention the aggression or show him your bandaged hand at this point, because, ironically, if people don't know a cat's dodgy history they behave in a relaxed, business-like and firm manner and the cat rarely resorts to violence.

√ Ignore the cat and do not approach it
The threat of a posturing cat is completely defused if it is ignored. It will sense there is no point in assuming the countenance of the devil himself if no one is watching. Occasionally the cat's aggression is motivated by its desire to manipulate the activity of its owner; if the owner then fails to respond to the behaviour it will soon stop. However, a word of warning: these cats will undoubtedly try harder before they give up (after all, it always used to work). For those interested in learning theory, this is referred to as 'extinction burst', and it may result in the cat temporarily appearing even meaner than before. This is why my next recommendation is to wear protective clothing; God forbid, but you may need it.

Many acts of aggression by cats are their way of saying 'Back off!' when owners or, worse still, complete strangers make unsolicited approaches to those unequipped to deal with the inexplicable behaviour of another species. The best strategy under these circumstances is to keep well away, thereby not allowing the situation to occur. Direct eye contact can also be a source of distress for some cats, so adopting a healthy air of disregard is the way forward. This takes the focus away from the cat and returns its sense of control of the environment and everything in it, which is such an important survival technique for the cat. Take away its sense of control and you may seriously compromise its quality of life.

√ Wear protective clothing

Anyone who has read any of my previous books about my work as a cat-behaviour counsellor will know that I appear to take a perverse pleasure in dressing my clients up in biker gear and all things leather. My intentions are honourable, however, as I am merely attempting to protect them from injury from their aggressive cats; leather boots and gloves are by far the strongest and most reliable guards against feline teeth and claws. I have even been known to resort to a motorcycle helmet to combat those cats that practise the terrifying 'launching at your face' technique.

The protective clothing is not only practical, it has great psychological benefits. Often owners who have been bitten or badly scratched by their cats behave very differently around them from that point onwards. They adopt a hesitant, nervous and jerky gait accompanied by rapid breathing and staring eyes that would make all but the most passive cat launch an attack. The provision of armour along with the knowledge that it will keep them from harm allows them to relax and move around the home in a far more natural way. This alone can have a great impact on those cats who have become a little too assertive for their own good.

√ Keep the cat out of the bedroom at night

This sounds like stating the obvious but you may (or may not) be surprised how many owners find it extremely difficult to deny their cat access to the bedroom at night. Often the cat has had an exemplary record of behaviour prior to 'losing it' and the owner finds it hard to understand why he or she should appear to punish a cat for one or two minor indiscretions.

If this sounds like your situation it's important to remember that probably, at the moment, you don't know why your beloved cat attacked you that fateful afternoon. We are not talking about a cat that has suddenly decided to turn to a life of crime; we are talking about a cat that is responding to what it perceives to be a genuine threat or danger. We may even be talking about a cat with a physical illness or a psychological issue that needs addressing. Actions should therefore be seen not as punishments but as potential remedies for resolving the problem. If a cat has attacked once there may be some stimulus associated with you that triggers the same emotional response and, therefore, another attack. If you are lying down in bed at night you are particularly vulnerable to damage to your eyes. Your cat isn't taking it out on you *per se*; he is as much a victim of this as you are. Protect yourself and your cat by shutting that bedroom door.

√ Asking your vet for a referral to a pet-behaviour counsellor
Apart from the above I would be very reluctant to recommend further self-help when confronted with an aggressive cat. There are so many potential motivations for this behaviour (and such varying levels) that it would be foolhardy to put human health at risk for any longer than you need to. Once you have visited your vet to rule out a medical cause for this problem you are in a good position to request a referral to a recognized cat-behaviour counsellor. You will then be given strategies to employ to modify your cat's behaviour while keeping yourself and your family safe.

Anxiety

Anxiety is quite a specific emotional state but it is possible that the cat hiding under the bed is actually exhibiting another equally distressing feeling. Your cat may be fearful: scared of other cats, loud noises or even you. He may possibly have a genuine phobia, driven to this dramatic response by his genes, his early experiences or a one-off event. Fears and phobias may look relatively similar, but the latter can be more difficult to treat as your cat could potentially be so scared that it will take him ages to learn the thing he is so frightened of won't actually kill him. Distinguishing between the two afflictions is probably something for the professional pet-behaviour counsellor, but in the meantime there are still a few dos and don'ts that will help.

Don't bother with:

X Thinking love will be enough

If your cat is genuinely anxious, then numerous attempts to cuddle, squeeze and kiss on the premise that 'love conquers all' will not work. Rather, they will further traumatize your cat and make the possibility of a good relationship in the future even more unlikely.

X Flooding his phobia

This is a term used for a technique employed in the human field to cure a patient of a phobia. The poor soul is exposed to the subject of his or her phobia (rats, heights, spiders, balloons, etc.) and then, with the aid of a cognitive therapist or similar professional, talked through the experience until such time as the heart and breathing rate has returned to normal and the person is no longer afraid. *Do not try this at home!* It works very well for humans, but candidates are carefully assessed to gauge their suitability for the process and they are guided through it before, during and after the exposure.

The problem with using this technique with cats is that they are cats. You can't reason with them or rationalize their fears verbally. They may look OK when you grab hold of them tightly for the fifth time, working on the principle that they will eventually enjoy it, but inside they are screaming.

X Pussyfooting around the house

You may think that you are being kind to your anxious cat when you tiptoe past and chastise your family for raising their voices or turning the television up. You may be well-intentioned, moving slowly and purposefully from one room to another to avoid startling him. If you could see a video of your behaviour you would realize how misguided this is. You look odd. You ooze an air of tension and disquiet that your cat will pick up on and he will think you are nervous too.

X Reassuring him

If you reassure your cat every time he jumps at the sound of the doorbell or a car outside or a dropped saucepan lid, he will think he was right to be scared.

Try instead:

√ Ignoring him

It seems like odd advice, but focusing on an anxious cat often makes him more anxious. If you ignore him your cat will feel that he is moving about invisibly, without drawing attention to himself. Anxious cats love the anonymity this brings and it restores the elusive sense of control that is so important to cats that feel permanently threatened. If he then chooses to approach you it would be perfectly acceptable to give him a brief and relaxed demonstration of your love with gentle words and a tickle on the chin.

√ Behaving normally indoors

If your cat is ever going to get used to the hubbub of family life you have to allow him to see it how it really is. Move normally, behave normally, and you will soon feel relaxed and contented and your cat will benefit from the calm signals.

✓ Using synthetic pheromones
Feliway is a synthetic version of natural feline pheromones, which are secreted from glands round the cat's face and head. They signal familiarity and security and have a calming effect. Feliway is available in two forms: a spray bottle and a plug-in device called a diffuser. The manufacturer's instructions are fairly comprehensive and give you a clear idea how to use this product to help tackle your cat's anxiety.

✓ Increased stimulation
Playing with your cat may seem like an indulgence with no more benefit than general amusement, but actually it serves a much more useful purpose. Play promotes exercise, both mental and physical, and helps build self-confidence. It creates a positive bond between owner and cat without all the hugs and cuddles that anxious cats find so distressing.

Excessive scratching

Scratching performs three important functions: it removes the outer husks of the front claws, exercises the muscles of the forelimbs and marks territory. However, scratching is also a behaviour that causes damage to wallpaper, furniture and carpets in the home.

Don't bother with:
X Throwing things or shouting
This is effective at the time but will only result in your cat scratching the same place in secret.

X Using aerosol repellents
These smell absolutely disgusting and will only be effective if they are used in the appropriate area repeatedly. By this time you will be unable to enter the room because of the smell and are therefore unable to use the very sofa you are trying to protect.

X Buying a scratching post that is too short
Many scratching posts are designed for kittens and they soon become unsuitable as the cat grows. A short one is never a substitute for an armchair or sofa.

Try instead:
✓ Providing the right number and type of scratching posts
This could mean the difference between a pristine three-piece suite or £1,000 down the drain. It is important to remember, however, that if a particular surface or object is being damaged you should provide an alternative that offers a similar experience: for example, if your cat is scratching textured wallpaper at a certain height, then the alternative scratching area should be vertical with similar texture and striations and positioned to allow him to stretch to the same level.

✓ Effective deterrents
Low-tack double-sided adhesive tape can be stuck over the area and this will provide an unpleasant, but not dangerous, experience when your cat next scratches there. (It will need replacing daily as the sticky surface soon loses its effectiveness when exposed to environmental dust.) This is particularly useful for sofas, chairs, carpets and some soft-wood furniture. It is essential to ensure that the tape is not too sticky (don't use carpet tape, for example!), since it could damage paws and fab-

ric. Commercially available double-sided adhesive sheets can be purchased from some household-cleaning suppliers specifically for this purpose. Tin foil can be used as an alternative to double-sided tape.

If wallpaper is being damaged, thin sheets of Perspex can be cut to size and fitted over the damaged area, using screws and rawlplugs if appropriate. Since it is smooth, this surface will be unattractive to scratch and it is also easily cleaned to remove any scent deposits. Double-sided adhesive tape could also be stuck over the affected area if the wallpaper is in any case so damaged as to require replacing.

Small vinyl pads (Soft Paws) can be glued over your cat's front claws by a veterinary surgeon and these will remain in place for six to ten weeks. Scratching will still happen, but damage will no longer occur and the cat can be retrained to use an acceptable area. This measure is advisable only for those cats kept exclusively indoors, as vinyl pads seriously compromise your cat's ability to defend himself or climb outside.

Synthetic feline facial pheromones (Feliway) sprayed over the area that's being damaged by scratching should deter future use.

House-soiling (inappropriate urination and/or defecation)

Unfortunately this problem is by far the most common issue facing cat owners. Stress, in a susceptible cat, seems to travel direct to its bladder or bowel and this can have disastrous consequences. House-soiling is often referred to as inappropriate urination or defecation; I have always thought that the word 'inappropriate' is a huge understatement. It isn't just inappropriate: for the household it is stress-promoting, soul-destroying, relationship-wrecking and incredibly expensive.

Cat urine, in particular, can cause thousands of pounds' worth of damage to carpets, curtains, walls, electrical items, tiling and furniture before owners realize the problem isn't going to go away on its own and help is needed. So, without further ado, take the following advice and tackle the situation with courage and enthusiasm. It can be resolved!

Don't bother with:

X Rubbing his nose in it

I can't emphasize enough the futility of this retaliatory gesture. Equally pointless is reasoning with him, shouting at him or bursting into tears (although feel free to do this in private rather than bottle up your angst). Your cat will misinterpret all of these acts of desperation as your response to the danger and threat he is currently feeling. The result of this is more soiling and certainly not a resolution. Many owners at this point would say 'He knows when he's been naughty so surely punishment works'. I would say that his low-slung body language or quick exit from the room is resulting from anticipation that you are going to 'lose the plot' again. Your cat will rarely associate your anger with his original act and, after all, it really wasn't a dirty protest.

X Smacking

I'm sure most owners would never resort to such heinous acts of violence, but there are still those who believe 'I was smacked as a child and it never did me any harm!' That may well be the case, but a cat is not a child and, even if it was, there are much better ways of dealing with this problem. Smacking is distressing for the animal and it will ruin your relationship.

X Bleaching the soiled areas
Bleach may disinfect the area sufficiently but it will break down to ammonia compounds not dissimilar to the urine you are trying to remove. The more bleach you use, the more likely your cat is to return to that area. See the 'Try instead' section for alternative strategies for cleaning.

X Smelly deterrents
Distributing orange peel and pepper in the targeted areas is pointless; it looks dreadful, ensures that anyone coming to the house knows you have a cat-pee problem and, if vaguely unpleasant for your cat, will only make him choose another corner of carpet to destroy.

X Other deterrents
Tin foil, newspaper, plastic sheeting and pine cones: none works in the long term. As I have always said, though, cats really are a law unto themselves. There are bound to be some cats out there that will never soil again once you employ one tiny sheet of tin foil, for example. Just in case your cat is one of those I should amend that last piece of advice and say, 'Try it once for a short period of time.' You never know.

X Adding soil to the litter tray
If you already have a litter tray indoors and it is this that your cat is shunning in favour of a small area behind the television, I wouldn't recommend trying to make the tray more appealing by adding soil from your garden. This is often quoted as being an attractive prompt to get your cat back to acceptable habits. I'm not averse to changing the substrate in the tray when cats start to pee or poo indoors, but soil wouldn't be my first choice. If you have ever tried maintaining soil as a litter material

indoors you will know that the smell that emanates from that after a few hours is ten times worse than anything coming from a piece of Axminster. There are better choices (see 'Try instead').

X Doing nothing

I'm a great believer in the 'wait and see' philosophy for many things but *not this*. If you do nothing you will lose your carpets, your friends and your sanity.

Try instead:

√ A visit to the veterinary practice

Even if the faeces and urine look relatively normal it is always important to rule out any pain, discomfort or illness that may be influencing your cat's rejection of its normal toilet habits. If your cat is suffering from cystitis, for example, it is highly likely that soiling will cease when the discomfort has gone.

√ Providing an indoor tray

If you discarded the indoor litter tray when your cat had free access outdoors, then it may be time to reintroduce it. When cats go outside they are mixing potentially with a large number of others. They have to be on their guard as territorial disputes are all part of the game. The less confident individuals are often reluctant to go to the toilet outside if they feel they are under threat from others creeping up on them when they least expect it.

If your cat is vigilant when he goes outside and seems to spend for ever deciding whether or not to venture forth, it could be because he is being bullied. He may even have had a bad experience while watering the daisies. In that case it is extremely unlikely that he's going to be keen to risk it again.

He will probably instead retain urine for as long as he can and then, unable to continue, relieve himself on your duvet. This represents the ultimate toilet: a yielding surface underfoot with a strong familiar scent that signals safety and security.

To tackle this problem I strongly recommend that you provide a permanent and discreetly located indoor toilet for emergencies. Unfortunately, since who wants to use an outdoor privy when there is a toilet in the comfort of the house, this will probably become his one and only latrine and the garden will be forgotten. You are destined to his lifetime of lucky dips with a litter scoop, but what price a pee-free duvet?

✓ Extra litter trays

If you have more than one cat this may be relevant to you. Multi-cat households can be fraught with tension (see Chapter 7 for further details) and often tactics used are covert and subtle. Crumbling your opponent by any means other than tooth and claw is a sensible survival strategy and cats have developed psychological warfare into a fine art. One very effective manoeuvre undertaken by the more assertive cat is litter-tray blocking. How better to send your opponent into a flat spin than by denying him the ability to use the toilet? It is hard to look mean and feel confident if you have a full bladder and bowel. These poor tortured souls will seek out any nook or cranny away from their adversary when they can eventually hold on no longer. (I'm sure you can now see why your punishment on top of that trauma is just too much to bear.)

There is a formula to this litter-tray provision; anyone who has read any of my books should know it off by heart. That formula is: ONE TRAY PER CAT PLUS ONE EXTRA IN DIFFERENT LOCATIONS. Therefore if you have three cats, for example, you would provide four trays in four different discreet locations. If this sounds like a lot of litter trays with nowhere to put them, it is probably wise to review the number of cats you are keeping. If there is a problem with inter-cat aggression in your house, and one of your cats is soiling as a result, you have to take action. If you can't bear the litter trays, then maybe one of your cats needs to leave home and live out his days in luxury as a single cat.

√ A second litter tray

If you have one cat and one tray you may still want to try the litter-tray formula. Providing a second tray in a discreet alternative location may satisfy even the most fastidious cat that has suddenly decided he absolutely refuses to pee and poo in the same place (Persians, Birmans and British Shorthairs, you know who you are!). He may also have had an unpleasant experience while using the original tray. I once saw a patient whose litter training went out of the window when he was bashed on the head mid-stream by a copy of the Yellow Pages coming through the letterbox. The moral of this story, of course, is: don't put the litter tray next to the front door.

√ Changing the litter substrate

It's ironic that often fastidious cats shun the most economical, lightweight, eco-friendly litters. What do they prefer? Yes, you've guessed it: the heaviest, messiest stuff that is the most difficult to dispose of, ending up tracked all over the house until the place looks like a weekend beach hut. It is the fine, clumping sand-like litters, often referred to as 'supreme', 'premium' or 'ultra'. These are all euphemisms for 'very expensive'. You can easily test this preference (if you are still not convinced) by placing a pile of builder's sand outside your house. Cats will come from miles around to make their deposits. I blame their ancestor, the African Wild Cat, for that one.

√ Removing any litter-tray liners
Tray liners are great in theory. You line the tray, fill with litter, the cat uses it and you discreetly fold the edges of the full liner and dispose of it as a hermetically sealed parcel of nastiness. However, cats have claws and their instinct is to attempt to cover their eliminations. The result is a torn, leaking liner that cannot be disposed of without a much more hands-on approach. The other down side is that when your cat gets his claws caught in the polythene he pulls them free, causing a shower of litter dust to fly into his eyes. As he sees it, this is a perfectly good reason to go behind the television instead.

√ Not using litter deodorants
This is another wonderful invention in theory. How lovely to have the fragrant scent of spring flowers coming from your cat's litter tray rather than the more usual stench of yesterday's Whiskas. This isn't quite so appealing for your cat, though. If you can smell it all the way up there with your nose, can you imagine how it comes across to an animal approaching it at head height with up to ten times more olfactory epithelium than a human? If you clean the litter tray regularly and effectively you do not need additional odours to mask the original. There should be no odour!

√ Cleaning the soiled areas
It is such a temptation to resort to bleach, the killer of 99.9 per cent of germs, but please resist. Instead, remove as much as possible of the liquid (in the case of urine-soiling), using kitchen paper towel, as soon as the indiscretion is discovered.

There are now a number of cleaners on the market specifically formulated to combat the stain and odour left by pet urine and faeces. I have my favourites and I certainly recommend you ask your vet for his or her personal experience of these products and which they would advise you to use. Alternatively, you could try a home-made cleaner:

1 Apply a solution of biological washing powder or liquid (one part powder to nine parts water) via a spray bottle on to the affected area (don't saturate the carpet as this merely pushes the urine further into the underlay and floor underneath, so keep the area as dry as possible throughout the process).
2 Wipe clean with paper towel.
3 Rinse with clean water, again applied with a spray bottle.
4 Wipe clean with paper towel.
5 Spray (yet another bottle) with *surgical* spirit (not white or methylated, please: that would be a very messy mistake).

This concoction may fade your carpet but at least your cat will be less inclined to return to the area in the future.

Don't forget that you will also need to take some of the other measures recommended here to get to the root cause of the problem.

√ Asking your vet for a referral to a cat-behaviour counsellor
If you have attempted the above methods and your cat continues to soil, then I recommend a referral. It will be such a relief to find someone who understands what you are going through. The moral support alone is worth its weight in gold.

Inter-cat aggression

In an ideal world all cats would love one another and never fight with their neighbours or threaten their brothers.

Unfortunately, life isn't perfect and we have always to remember that cats are territorial creatures. We can try to breed aggression out of them or live in a permanent state of denial, but, at the end of the day, truth will out. It is unreasonable for us to presume that cats – even those living in the same household – won't fight. If cats have been brought up together or are part of the same family, then the likelihood of problems occurring diminishes. If you are very lucky you can, under the right circumstances, put a motley crew of half a dozen cats together with no familial connection and you will have peace and harmony. It really is the luck of the draw.

If you are one of the unfortunate owners who is experiencing cat tension within your home, you will know how hard this is to live with. No matter how many times people like me tell you 'It's perfectly normal behaviour' it isn't fun if your cats argue all the time or, worse still, intimidate the nervous one of the group into a quivering heap.

Don't bother with:

X Getting uptight

It's normal for owners to worry when cats fight; after all, if your children were doing it you wouldn't like all that noise and violence. However, cats really are different and resolution of disputes is not achieved by sitting round a table, discussing the problem and reaching a compromise. It's about devil-cat posturing and blood-curdling screams; if you are very lucky the other cat will retreat at the sight of such a monster or, if not, there will be a blur of teeth and claws and a victor will emerge when the fur settles. Unfortunately, if you still can't grasp that this is just what cats do, you will probably get rather tense when it occurs. If your cats are failing to get on most or all of the time it is likely that this tension will remain in you likewise.

Ironically, this will simply make your cats more uptight and potentially escalate the aggression between them.

X Using arms or legs to break up a fight
This is dangerous in the extreme. In the heat of the moment it is difficult for your cats to establish where the other cat finishes and you begin, so you are likely to get hurt.

X Dividing your house and separating the warring cats
This often sounds like the perfect compromise. The cats don't get on, you don't want to lose either of them, so you give one cat the run of the kitchen, living room and dining room and the other the bedrooms and hallway. There are notices everywhere saying KEEP THIS DOOR SHUT AT ALL TIMES and you get used to eating supper on your bed or sleeping on the sofa because you are trying to timeshare your affection but it's gone all out of kilter. You could argue that, if you are prepared to accept the inconvenience, the cats enjoy the perfect compromise. This is, I'm afraid, rather misguided. Your cats now know that their nemesis is just one door away. They want to deal with the conflict that is present between them but they are unable to do so. They feel the danger but cannot seek out the adversary to defend their territory. This is probably worse than sharing the home without the segregation in the first place; at least then the target was present and tangible.

X Putting them both in cages and placing them side by side to get used to each other
I understand the principle behind this (I think), but the method is as likely to fail as two people, each holding a cat, approaching each other, oblivious to the risk of injury or escape. Cats need to know they can escape danger and both

methods will teach the cats much more about the nature of captivity and restraint than it will positive things about each other.

X Shutting them in a room and letting them get on with it
Pupils of the School of Tough Love often advocate this method as 'It's worked for me and I've had cats for over fifty years'. For so many reasons I would recommend that you don't do this. I am sure there are many people out there who did this and then thought that the cats were fine with each other thereafter. Some cats probably would agree to disagree (a case of luck triumphing over judgement), but others would be deeply distressed by the whole thing and develop all sorts of strategies to cope with the situation: leave home at the first opportunity, keep out of trouble and hide a lot, eat, eat, eat, etc.

Try instead:
✓ Keeping calm
This will ensure you are not a contributor to the tension in the atmosphere.

✓ Choosing siblings or cats with sociable parents
You are less likely to have problems if you bring cats up together, particularly those from the same family. Sociability is inherited to a degree, so see how the parents behave towards other adult cats before picking one of their kittens.

✓ Separating fighting cats with distraction or pillows
The less intense fighting (or the behaviour leading up to a rumble) can sometimes be stopped with the distraction of a loud noise. A handclap or shout will often do the trick. If the fighting looks so intense that at least one of the cats is going to

get seriously hurt, you can use large pillows or blankets to separate them and give you the opportunity to remove one of them from the room to calm down.

√ Thinking carefully before getting another cat

It's really important to establish whether or not your household can take another cat before you embark on the acquisition process. You need to look not only at the number of cats in your home but also the number of cats outside. If you are in a neighbourhood with a dense cat population you may just be adding to the stress of your existing resident cat or cats. If you already have a multi-cat household there is always the consideration that, if harmony reigns at the moment, why risk spoiling it by crossing the 'one cat too many' threshold.

√ Introducing new cats carefully

If you have made the decision to go ahead, then the introduction should be done gradually to ensure the best possible results. This may be time-consuming and frustrating if you are an impatient soul, but it's worth it in the end. The principle of a successful cat introduction is to do it in stages:

1 Introduce the scent of the newcomer to the resident cat and vice versa.
2 Allow the two to observe each other without the possibility of physical contact.
3 Allow them physical contact.

Each stage is started once the previous one has achieved a satisfactory result. The whole process can take anything from days to months, depending on the circumstances and the personalities involved. Chapter 7 includes a comprehensive step-by-step guide for this technique.

√ Providing sufficient resources to prevent competition

Aggression will escalate if cats feel there is a scarcity of those things they consider are important. These are referred to as 'resources' and they include feeding areas, water bowls, scratching posts, beds, high perches, private areas, litter trays and toys. Other things can become important to individuals, so it's worth observing your cats to see where or when the tension appears worse. The formula of 'one per cat plus one extra in different locations' works for all resources, so if you can put up with the clutter this will certainly help.

√ Not getting involved

This really is a cat thing and it's impossible to know exactly what is being 'said' between them. If you are this much in the dark about what's going on it just isn't possible to intervene with the appropriate input. Many owners think that the answer is to give both cats equal attention, even to the extent of stroking one while stroking the other. I honestly don't agree; I think it is far better to allow the cats to decide who has access to you and when. Owners often become a very important resource, the provider of food, security, entertainment and all things in between. This creates competition between cats in the household as they compete to get access to such a multi-functional asset. If you can bear to, back off a little. Stop being the instigator of mealtimes: put dry food down in the morning for feeding throughout the day; stop being the doorman: fit a couple of cat flaps giving access to two different areas; stop being the chief entertainer: provide toys that the cats can amuse themselves with. All of these strategies will defuse tension between your cats rather than make them think you don't love them any more.

Over-grooming

This covers a multitude of different manifestations under one umbrella. Over-grooming can describe:

- licking areas excessively, causing breakage or removal of hair and subsequent alopecia (baldness)
- plucking areas excessively, causing similar damage
- licking, plucking and chewing so severely that skin is broken, becoming infected and severely traumatized
- chewing the tail, causing extensive damage, possibly requiring amputation
- chewing limbs, causing trauma to skin and underlying tissue
- excessive scratching around the head and neck, causing severe alopecia and ulcerated lesions

This list is not exclusive; there are many different ways for a cat to make a mess of himself. Some over-grooming cases are merely cosmetic. The cat has bald bits, often symmetrical, and you may find, in time, that the problem resolves on its own or waxes and wanes throughout the seasons. There is a school of thought that says the investigation and possible treatment of this problem may be worse than having a few bald bits. However, I would never recommend you view these things lightly. Over-grooming is always a sign of something, so even if the symptoms are mild it's worth monitoring. Obviously if the skin is damaged or your cat appears distressed, there is no justification for delay in taking the necessary action.

Don't bother with:

X Automatically presuming it's 'stress'

When confronted with a partially bald cat many vets, as well as owners, are too quick to jump to the diagnosis of 'stress-related' alopecia. There are many cases where stress has been a trigger or an accelerant of symptoms, but genuine psychogenic alopecia is rare. This problem needs investigating by your vet, as there is likely to be a physical cause.

X Punishing your cat
If your cat is easily interrupted when he over-grooms, then don't be tempted to punish him, shout or generally distract him with loud noises in the belief that this will break the habit. He will certainly think twice before doing it in front of you again and become, instead, a closet licker, doing it under cover of darkness or when you are out.

X Knitting him something fetching
Some owners, to deter further trauma to their cat's skin or even to provide warmth over the bald bits, will knit or construct vest-like clothing that covers the affected area and denies further access for the cat. Just like bandaging a traumatized tail, it draws constant attention to the area (as the cat madly attempts to remove the offending item) and creates stress even if it wasn't there in the first place.

Try instead:
✓ Effective flea control
Many cases of over-grooming relate to flea saliva hypersensitivity. Treatment includes both topical application to the cat (usually in liquid form, placed on the skin on the back of the neck) and environmental treatment to destroy or arrest the development of any of the stages of the flea (from egg to larva to pupa) that may be taking place in your carpet. Nobody likes to admit they may have a flea problem in the home, but any cat, no matter how beautiful, can pick up a flea from outside and inadvertently bring it indoors. Even *you* can bring an oppor-

tunistic flea home on your clothing. Modern flea treatments will kill the little biters on contact with your cat's skin and fur (rather than acting only after the fleas have bitten and thereby ingested the chemical) which avoids further attacks of itching.

√ Consulting your vet
If your flea control is good but the problem continues, then the next step, rather than further self-help, is a trip to your vet. He or she will examine your cat thoroughly and carry out various tests to ascertain what, if anything, your cat is allergic to that may make him itch. If the fur loss is restricted to a specific area your vet may suspect there's a pain component to the problem, so don't be surprised if X-rays are recommended or your cat is given pain-relieving medication. You may even find that your cat is put on an exclusion diet – specially formulated food that is a novel source of protein and carbohydrate – to identify any possible food allergies. Only if all investigations, including a referral to a dermatologist, show nothing abnormal should you consider consulting a behaviour counsellor.

Pica

Pica refers to the consumption of non-nutritious materials. Favoured delicacies for the susceptible individual include wool, rubber, plastic, leather and cardboard. If you are very lucky your cat will stop short of actually swallowing the stuff and remain perfectly satisfied with just causing hundreds of pounds' worth of damage. If you are not so lucky the cat will not only cause hundreds of pounds' worth of damage but also cost you hundreds of pounds for surgery as your vet removes the offending material from his intestines. This is not a cheap problem.

Don't bother with:

X Punishment

As always, this just doesn't work; to the seriously 'alternative' cat psyche this may be highly rewarding. Shouting and screaming when yet another cashmere sweater is consumed is understandable but likely to become an amusing part of the game for the cat.

Try instead:

√ Lacing attractive items with obnoxious flavours

I could justifiably have put this in the 'Don't bother with' section, but not all pica habits are as intractable as most. For those cats that idly chew rather than actively seek-out-and-consume-in-the-blink-of-an-eye, you may find the disgusting taste of Olbas Oil (eucalyptus oil) or Bitter Apple (used to deter animals from chewing at surgical stitches) is sufficient to deter them from this bothersome habit. However, if your cat eats it despite the foul taste you are probably fighting a losing battle.

√ Increased stimulation

Many cats that develop this habit maintain it into adulthood if their existence falls short of the more natural 'hunting, shooting, fishing' lifestyle. A cat with little or no access outdoors needs a great deal of stimulation indoors – things to do that are more rewarding than consuming leather, for example. He may feel under pressure from the other cats in the household and unable to put enough personal space between them and himself to escape persecution. Under those circumstances, it's hardly surprising that he resorts to hiding under the bed and chewing a pair of sling-backs. The ideal solution is free access outdoors; if this isn't possible, then a secured garden or outside pen is a second-best option.

√ Changing the diet

This is something to consider, particularly if your vet or a behaviour specialist is assisting you to manage this problem. Diets high in fibre that make your cat feel full may be helpful in reducing the motivation to perform the misdeed, particularly if the items are invariably swallowed once they are chewed sufficiently. Your vet will be able to recommend a veterinary formulated diet; don't try a home-made version, as extra vitamins and minerals need to be added to maintain coat and skin condition and general wellbeing. Fibre can cause the gut to work more quickly than normal and this can lead to a lack of absorption of essential nutrients in the food. A cooked knucklebone of lamb or beef with some residual gristle and meat on it can be attractive to some with a more logical choice of inappropriate 'food' items. However, in my experience cats that consume plastic and other man-made materials are not automatically drawn to something so obviously normal in the dietary line.

With luck you may be able to direct the habit on to something safe to chew that is pretty well indestructible, such as a dog hide chew. The smaller version is the best (about 5–6 inches long and ½–¾ inch in diameter), but they do need to be soaked first in hot water to make them remotely attractive to a texture-loving pica sufferer. Putting on a few drops of fish sauce or similar strong flavour may attract your cat in the first instance. A green wood twig may also be appealing, but it's always worth checking that it isn't from a poisonous tree before offering it as a chew stick.

√ Consulting a cat-behaviour counsellor

Let's hope your cat will grow out of this problem by the time he matures, from the age of eighteen months onwards. If he is kept indoors or with other cats, though, he may retain this behaviour into adulthood. There are rarely any guaranteed cures, but a specialist would help you to manage this condition better. Your veterinary surgeon may prescribe a tricyclic anti-depressant (for the cat, not you) that will work alongside the behaviour therapy to 'retrain' your cat's brain. Drugs such as Clomicalm (clomipramine hydrochloride) or Prozac (fluoxetine) are frequently used, with variable effects. Despite my reservations about the use of potent drugs on our pet cats, I think it is definitely worth a try, as this problem has a profound effect on your lifestyle as well as the cat's and it is very hard to live 'fabric-free' and almost impossible to deny your cat access to all the variety of tasty treats. I have visited owners with pica-suffering cats and on one occasion a small piece of my trouser leg disappeared into a determined Siamese cat as I was taking notes. It really can happen that quickly.

Territorial aggression

This is specifically concerned with those disputes your cat enters into outside your property and, potentially, in someone else's. Some cats take defence of their territory very seriously and are not content to settle for a bit of staring and growling and fluffing-up. Your Tigger may be a pussycat indoors but it is entirely possible he is a tiger when confronted with Mrs Smith's Sooty at No. 8. If Tigger attacked Sooty while the latter was menacingly approaching your back door it would be impossible for Mrs Smith to complain. Unfortunately cats like Tigger attack cats like Sooty on the victim's home ground, and this is one violation too many for most. Once again, it is just a case of cats behaving like cats, but there are strategies you can employ to ensure you do not become *persona non grata* in the neighbourhood. (I have discussed this already in Chapter 6 but it's worth repeating).

Don't bother with:
X Getting involved in disputes with your neighbours
No matter how reasonable you are being, as far as your neighbours are concerned you own a monster of a cat that is hooked on violence. The reality of the situation is that cats are territorial creatures and it is normal for them to fight and be aggressive. The level of fighting and aggression considered to be 'normal' is a debatable point. Your neighbours will never see their cats as anything but victims and they will expect you to do something about it. Neither accept nor deny culpability, but suggest possible solutions (see 'Try instead' below).

X A constant state of denial
I'm afraid this is your cat they are talking about, no matter how cute and cuddly he is at home.

Try instead:
✓ Removing the victim's cat flap
Many (not all) territorial cats are attracted to another's cat flap; it's an easy point of entry to a nice place with food and a feeble cat to beat up. Once the poor unsuspecting victim has been attacked once in the 'safety' of his own home, it will be almost impossible for him not to watch the cat flap day and night, expecting round two. He may well decide that going outside isn't much fun anyway, even if he did before, so the removal of the flap will not necessarily blight his life. On the contrary, it may give him the renewed sense of security he craves.

If the neighbours are reluctant to provide the litter tray made necessary by the enforced confinement, they are going to have to escort their cat outside regularly for supervised and protected toileting (it's surely far easier to get the tray installed). If they flatly refuse to remove the cat flap, you could suggest assisting them in the purchase of an infrared exclusive-entry flap. If they refuse to place a collar on their cat to accommodate the necessary battery (now they're just being awkward!), you can recommend the latest state-of-the-art cat flap operated by the identification microchip in the back of their cat's neck; this way no battery or collar is required.

√ Timeshare

If you have a reasonably good idea of the extent of your cat's territorial disputes you could suggest a system of timeshare – for example, your cat goes out at a time when all the other neighbours have promised to keep their cats in and vice versa. This is worth a try, although in my experience cats with a strong territorial instinct don't like being confined indoors against their will. They also have extremely good body clocks and soon learn to anticipate when you are going to shut them away. Then their failing to return home at all soon puts a stop to that plan.

√ Securing your garden

If the problem becomes really nasty – for example, neighbours are asking for reimbursement for vet bills or making vague threats – you could consider securing your garden (see Chapter 6) and allowing your cat access to the house and this area only. You would have to ensure there was sufficient entertainment within that area to stimulate such an active cat. Once again, the nature of this beast will make him the least adaptable to having his movements curtailed in this way. This could potentially be extremely frustrating for him, as he senses the presence of other cats but is unable to 'service' his territory accordingly.

✓ Re-homing

Believe me, this is better than finding he's gone 'missing' one day, as has happened in a couple of cases I've been involved in over the years. Most owners will have a much more measured response to this problem than re-homing, but in certain instances desperation takes over. In any case, it could be that your cat would prefer a lifestyle without all these enforced restrictions on him. An open rural environment with plenty of scope to carve out an expansive territory, full of cats with a little more 'oomph' to fight back every now and then: paradise!

Urine-spraying

Urine-spraying is a perfectly normal marking behaviour for cats; it gives a visual and olfactory signal to others and reassures the sprayer in areas of conflict within the territory. Males and females, whether neutered or entire, are capable of spraying. This is fine when it is a technique employed to anoint bushes or fences but a disaster when it's your lovely velvet curtains.

Don't bother with:

X Rubbing his nose in it

Once again, this is an action employed by those who adhere to the 'school of dominance and punishment' method of teaching, or by someone who has misguidedly listened to an 'animal lover' with a lifetime's experience of cats and dogs who seems to know best. Well, it is possible to spend a lot of time with cats and still get it wrong, so please don't follow this advice if you want a good relationship with your cat at the end of it. Punishing a cat for an act performed some time ago, or even one that from the cat's point of view was unavoidable under the circumstances, will distress him further; it's also likely to lead him to spray even more frequently, as you have increased the sense of threat within the home.

X Punishment
See above: exactly the same thing applies. You may be successful, with a handclap or a shout, in deterring your cat from spraying on your toaster when you are slicing bread beside it, but he's sure to be back to do it when you are not there.

X Reasoning with him
Now, some of you may have the sort of relationship with your cat that seems to operate on a higher level. You can say 'Please, don't put your paws on the table' or 'Come to Mummy for a cuddle' and your obedient little cat, understanding every syllable, will behave accordingly. However, I have to say, and I know that you know deep down this is true, your cat doesn't really speak English. The tone of your first request, for example, may have interrupted the behaviour or he may have had no intention of putting his paws on the table anyway. Your second request will have been heard as an invitation for contact, just like any other jumble of phrases you may care to use in the same gooey voice. So, sitting your cat down and explaining why the behaviour is unacceptable, and even evoking the 'naughty step' technique of child management, will have no consequence whatsoever. You may think I am being flippant in even suggesting that anyone would consider it

useful to 'reason' with a cat, but I have spoken to and witnessed many owners doing exactly this. There is a positive side: the voice used is often soft, firm but loving. This is what your cat will hear and, providing you are not holding him firmly by the shoulders or wagging your finger at him at the same time, you won't do any harm.

X Orange peel
Citrus peel is often cited as a good scent deterrent for urine-spraying and other 'indiscretions'. If you put grapefruit, orange, lemon or lime peel on your carpet you will, at best, deter your cat from spraying in exactly that location. If you keep putting more peel down every time you find a newly anointed spot you will end up looking like a very messy vitamin C junky.

X Covering your entire house with Bacofoil
As above, although I'm not quite sure what having tin foil everywhere would say about you.

X Plugging in fifteen different air fresheners that puff out the scent of everything from apple to zabaglione at ten-second intervals
I am constantly amazed by the variety of smells you can emit into your home. Depending on the source, you can burn it, plug it in or battery-operate it. If you want your home to smell like new-mown hay or freshly baked bread there is almost certainly an air freshener for you. However, if you fill one room with apple fragrance and another with magnolia, you will undoubtedly smell like someone who is trying to mask the smell of urine. Manufacturers don't produce a room fragrance that smells of ammonia, but unfortunately this nevertheless is what your home will smell like. The double whammy to this

is that the conflict of the additional strong smells could potentially deeply distress and confuse your cat, making the problem worse, as most of these contraptions are designed to pump the fancy odours from wall sockets directly into his face.

Try instead:

√ A visit to the vet to rule out urinary tract disease
Just to confuse matters, cats that spray urine may not be marking but suffering from uncomfortable symptoms associated with urinary tract disease. This would be particularly relevant if your cat is spraying near the litter tray or passing large quantities while doing so.

√ Cleaning the sprayed areas with hot water and kitchen paper towel, then applying Feliway spray
Feliway spray is a product that I mention frequently in all my books. It contains a synthetic version of a part of the feline facial pheromones common to all domestic cats and secreted from glands around the head. These pheromones are deposited as part of the normal marking behaviour when cats rub their faces on items of furniture or doorways throughout the house to provide themselves with reassurance and a sense of familiarity. Research has shown that cats are reluctant to spray urine where they detect the presence of these pheromones. Feliway is a great product as an adjunct to behaviour-modification techniques, but it can be really stretched if it is used on its own. It is possible that this may merely redirect the behaviour elsewhere and be only marginally more effective than orange peel and tin foil. However, it's worth a try if you are going to do a bit more DIY behaviour therapy by following the advice throughout this particular section.

√ Increasing interactive play activity

Play tends to be a leisure activity for cats and they are rarely quick to play if something is bothering them – particularly if that something is other cats. If your cat can be persuaded to indulge in a little feather-and-fur fun, it can have an incredibly positive impact on his emotional wellbeing. Be warned, though, if your cat is spraying because of an issue with another cat in the household you will be wasting your time flinging yourself about the living room with a piece of string in tow; he will flatly refuse to play if the other cat is about, so you will have to occupy his nemesis elsewhere.

√ Placing small bowls of dry food at sprayed sites

This tip is all about changing the scent profile of the sprayed area to something a little more positive; again, research has shown that cats rarely spray in an area they perceive as a regular feeding station. Unfortunately many cats have not read the book 'How to be a cat' and they will gladly either eat the food first, then spray the wall behind as usual, or spray anyway, splattering the food and rendering it inedible. It's worth a try, though.

√ Increasing resources within the home to a total of one per cat plus one

If the source of the tension your cat is experiencing is a dispute in a multi-cat group inside the home, you can do a quick recce to see if you have enough 'cat stuff' to prevent unnecessary competition. A single feeding area can be a popular arena for fisticuffs, particularly if you have just two set mealtimes a day. Under these circumstances food looks scarce and cats have to make sure they get there first. The things that cats fight over include resources such as feeding bowls, water bowls, scratching posts, beds, private areas, high resting places, toys and you. If you have the same number of these (plus one extra) as you have cats, there is less likely to be competition. The fundamental flaw in this formula is that you can't replicate yourself.

√ Reviewing litter-tray provisions
The most important resource for cats that live indoors or have restricted access outside is the toilet. For example, if Sooty, Ginger and Tigger have only one litter tray between them, there is bound to be trouble. Sooty, the more assertive in this case, will spend half his waking day scraping about in it and the other half guarding it, so that Ginger and Tigger have permanently crossed legs. This may sound comical but it could lead to permanent health damage and painful cases of urinary tract disease in those cats too frightened to pee. The 'one per cat plus one' formula derives from the standard advice for litter trays. The most important part of the equation, for all resources but especially for litter trays, is placing them in different locations. It just isn't good enough in a four-cat household to have five trays lined up side by side in the utility room. The wily cat blocks them all simply by sitting casually in the doorway.

Training your cat

Not all cats have behavioural problems and some never commit a major misdemeanour. That doesn't mean, however, that all cats behave as we would wish them to all the time. We don't want the cat stealing the pizza topping from under our noses (I saw this done once by a clever old cat called Hilda; she got the cheese and the horrified guest got the base in a manoeuvre that took a split second). We may have hygiene issues with the cat walking across the work surface in the kitchen, or even have some general house rules that we wish our pets to adhere to. I have always been against training cats in principle, often remarking that if you are that keen to do so maybe you should have bought a dog instead. However, that's not the whole picture and there are often occasions when life for cat and human would be a whole lot easier if, once in a while, cats did what they were told.

House rules are not a bad thing, bearing in mind that:

- You are not being cruel if you eat your own food yourself without sharing it with the cat.
- You do not have to give your cat something every time you open the fridge door, despite the fact that your cat insists you should.
- Your cat can be prevented from doing something if it causes you a problem financially, practically or emotionally. Remember, he does not pay the mortgage.

Training a cat is fairly straightforward providing your instructions are clear and you reward the right behaviour at the right time. Cats are not small dogs; they rarely if ever perform for praise alone. If your cat is motivated by a particular food it is possible to create a link in his brain between, for example, hearing the sound of a particular word and the sound of a treat. We see this all the time when rattling the cat biscuit box, opening the fridge or getting the tin opener from the cupboard. When I lived in Cornwall with my seven cats they were always out and about in various places at mealtimes. I therefore 'trained' them all to come to the one communal word '*Chicken!*' rather than go through a roll call at the back door of seven separate and rather daft names. It never failed to work.

Training your cat to get off the table

There are some occasions when training is not effective. We all know that the constant shout of *'No!'*, accompanied by the removal of your cat from the table or work surface, will only result in his walking on it when you are not there instead. Successful training (whether to reward or deter behaviour) relies on a consistent and persistent pattern of consequence to a particular action. Your cat will learn not to do something if the consequence of his action is unpleasant every time he does it. It's also important that the bad thing that happens isn't associated with you. This is sometimes referred to as 'Act of God' or remote punishment; here is an example:

- Create a paper template of the surface you wish your cat to stay off.
- Remove all items and attach the paper to the surface using Blu-Tack (or similar low-tack adhesive method) to prevent damage.
- Attach double-sided adhesive tape to the top of this paper in a criss-cross pattern, removing the backing sheet to leave the sticky side uppermost.
- Replace adhesive tape frequently to maintain its stickiness.
- Your cat will jump on to the surface, feel an unpleasant sensation underfoot and jump off.
- Provide an alternative location for your cat to jump up to that will offer a similar high vantage point for resting and general observation of his territory.

How to deter begging

You can easily prevent your cat from begging at the table by starting as you mean to go on. If you never allow your cat to receive food from the table he will soon get the message and not waste time looking cute if there is nothing in it for him.

If he has got into the habit of begging (because you succumb to his charms and give him a little of just about everything from curry to crisps), then you need to work harder and show great determination to stop the habit once and for all. The biggest hurdle to overcome when you stop rewarding previously successful behaviour is the huge surge of frustration that your cat will experience. Here is the sequence of events:

- The cat begs at the table.
- The owner completely ignores (no eye, verbal or physical contact).
- The cat cries loudly and paws at owner's leg.
- The owner continues to ignore as above.
- The cat jumps on table and attempts to steal food.
- The owner gently but firmly lifts cat from the table (without verbal communication) and places him outside the room.
- The cat scratches at door and howls.
- The owner continues to ignore as above.

Once this has happened several times (perseverance is the key) the cat will eventually give up. An even easier way to achieve the same result is to shut the cat out of the room at mealtimes and feed him cat food in his own bowl at the same time. One very important thing to remember is the significance of 'giving in just the once'. If after several attempts at ignoring the begging behaviour you relent and pass a small amount of chicken under the table, it will be virtually impossible to stop the behaviour in the future. You have now appealed to the slot-machine gambler in your cat and the lure of an intermittent, random reinforcement will be too much to resist.

Training your cat to answer to his name

This will take place when the stimulus (your shouting his name) is associated with the consequence (presence of food). Your cat will sometimes learn to come to his name by accident, but putting the right effort in at the beginning will ensure his attention.

- Decide on the cat's name and stick to it; changing a cat's name will only confuse the issue.
- If you have more than one cat, then pick a communal word/command to call all the cats en masse.
- Names (or command words) should be said in the same way each time – for example, in a high pitch with a different tone for each syllable. Single-syllable names can be said twice (e.g., Puss Puss).
- Reward the arrival of your cat with praise and food and continued use of his name.
- Do not use his name in the same way when angry! The intonation of your voice when you are mad at him should carry a different association.
- Further reinforce the behaviour once it has been established by calling your cat every now and then without providing food (see above re gambling-addicted cats).

Clicker training

I recently was fortunate enough to visit London Zoo for a 'behind the scenes' look at some of the residents. One experience that I shall never forget involved watching an enormous male lion called Lucifer being clicker-trained. Veterinary examination of wild animals is fraught with danger, but Lucifer had been trained, with the aid of a device I shall explain shortly, to show his paws, open his mouth and roll over to allow visual examination of most body parts without resorting to sedation.

The device was a small plastic box with a metal insert that, when pressed, emits a distinct sound. This is used as an aid to train an animal to behave in a certain way 'on command'. Basically the trainer waits for the animal to perform certain behaviour, then immediately this happens the clicker is used and a treat given. This is an effective way of ensuring the animal is in no doubt about what action was actually rewarded with the treat and takes all the vagueness and guesswork out of training.

In order to start the basics you can use either a commercial clicker or anything around the house that makes a sharp clicking sound – for example, a ballpoint pen and a stick that you can use as a target.

- Start the training session when you know your cat is hungry and have ready a dozen or so small biscuits that your cat loves.
- Take one of the treats and offer it to him, simultaneously clicking the ballpoint pen or clicker.
- Do this several times without verbal or physical contact so that he associates the click sound with the provision of his favourite treat.
- Don't worry if he gets bored and walks away. Try again tomorrow.
- Once your cat has formed an association between the click sound and the treat you can start shaping behaviour.
- Offer the target stick to your cat. His instinct will be to touch his nose to the end of the stick if you offer it to him. Once his nose touches the stick, click and treat simultaneously.

- Try this again, only move the stick slightly further away. Once touched, click and treat immediately.
- Allow your cat to walk away at any time if he is bored or full up. Try again tomorrow.

You will soon have a cat that responds to the stick and you can achieve a great deal by offering the stick in front of him and 'luring' him to desired locations, where he will get his treat.

Once you have established the pattern of action-and-reward you can teach your cat many other useful behaviours, without using a target stick, such as:

- acceptance of being handled or groomed
- recall
- alternative behaviour to anything undesirable, such as fighting
- staying in a particular place
- walking on a harness

There are various important rules to remember before you embark on creating the perfect cat. The drill won't work if your cat has food available all the time: he just won't ever be hungry enough. The timing of the click and treat are critical; get it wrong and you may reward the wrong behaviour.

CHAPTER 9

Middle-Age Spread and the Twilight Years

THE OLDER CAT

THERE SEEMS TO BE A PERIOD IN OUR CATS' LIVES WHEN THEY are suspended in a persistent state of youthfulness. How many times have you been asked 'How old is he now?' and you've replied, for example, 'Five', only to realize later, having checked the dates, that he was nine last birthday!

It's vaguely comforting to be in denial of our cats' advancing years, but I think it's a far better strategy ultimately to accept and embrace the ageing process. Although I'm not for one moment suggesting we anticipate our loved ones' demise, a degree of gentle monitoring during various stages will undoubtedly be beneficial and flag up potential problems, enabling them to be tackled early.

Probably the easiest way to do this is during the annual celebration of your cat's birthday. After blowing out the candles, you can take time to reflect on his age: whether any lifestyle, weight or behaviour changes have occurred over the past year. This will give you an ideal opportunity to establish that, for example, he's:

- looking a little rounder in the tummy
- not going outdoors so much
- getting lazy

We are really looking for behavioural, rather than physical, signs here but it may be a good opportunity to consider both now. Your vet will be assessing your cat's general health every year when your cat has his annual health check and/or vaccination visit, so this will be monitored anyway. Having said that, your vet will only be able to establish potential problems via physical examination and relevant tests. You are in a much better position to identify lifestyle issues that may be relevant, because you live with your cat every day and know his habits and fancies.

How old is your cat?

First of all it's important to get age into perspective and the best way to do that is to equate your cat's age with the human equivalent. The first two years of your cat's life correspond with a huge amount of human emotional and physical development, the first year being equivalent to fifteen years of human life and the second year to twenty-four.

Probably the best formula to use, for cats over the age of two, is:

Equivalent human age = 24 + [(your cat's age – 2) × 4]

If that confuses the less mathematically minded among us, here is a chart.

CAT AGE	EQUIVALENT IN HUMAN YEARS	CAT AGE	EQUIVALENT IN HUMAN YEARS
1	15	13	68
2	24	14	72
3	28	15	76
4	32	16	80
5	36	17	84
6	40	18	88
7	44	19	92
8	48	20	96
9	52	21	100
10	56	22	104
11	60	23	108
12	64	24	112

It's struck me as rather sweet that our cats, on reaching the age of twenty-one, would qualify for a telegram from the Queen (if there still were such things). And, for all those people who think I have been hugely optimistic in providing a chart that goes up to the age of twenty-four, I should say that I remember with some affection Stevie, the oldest participant in my Elderly Cat Survey, at the grand age of 26! It all goes to show what a huge impact diet, veterinary care and housing have on longevity in the pampered domestic pet when you consider that the feral tomcat has an average life-span of five years.

There has been a great deal of debate recently, particularly in the pet-food sector, regarding the appropriate terminology for the various life stages. Only a few years ago we all recognized and accepted 'kitten', 'adult' and 'senior' as being fitting but these terms don't seem detailed enough any more. The feline population (not unlike the human one) is getting older; you will now see reference to 'kitten' and 'junior' to cover the first two years of life, 'prime' to take the cat to six years, 'mature' up to ten years and 'senior' to fourteen. This leaves the rather unflattering term 'geriatric' for all those cats fourteen and over. Considering the accepted threshold for this was twelve only ten years ago, it just shows how rapidly we are progressing.

Assess his lifestyle

Maybe your cat is no longer a kitten yet he isn't old; he's just rumbling on, living alongside you and your hectic schedule. When every year, as suggested, you look at him with fresh eyes (I'd hate it to be a shock when you wake up one morning and suddenly realize he got old), remember that a major contributor to his overall health and vitality is his lifestyle. Ask yourself these questions:

- Is he playing quite so readily these days?
- Am I encouraging him to play as much as I used to?
- Is he going outside as often as he did?
- Is he a fair-weather cat, going outside just to lie in the sun?
- Does he get excited when I bring new things into the house?
- Are there more cats outside than there used to be?
- Is he as affectionate as he used to be?
- Do I pay him as much attention?

This isn't a definitive list and there are no right or wrong answers. How your cat develops as a unique personality and how he responds to his world will depend on his genes and his early experiences, but the things he needs to react to can be influenced by you to a certain extent to ensure he gets the best out of life. Unfortunately, if cats feel threatened by something they will often 'shut down' behaviour rather than actively show their feelings. Your cat may look utterly content as he sleeps on the sofa all day and on the bed all night, but this could indicate that he's too scared to do anything else. Cats fill voids created by their inability to do stuff with safe predictable behaviour. You can't get much safer than sleeping (or, often, feigning it), eating or grooming. It is no coincidence that many stressed cats either over-eat, over-sleep or over-groom.

If you want to have a healthy cat, in mind and body, then he needs to eat well (not to excess), exercise daily and have a life without chronic stress. In other words, he needs just what the average human does to live to a ripe old age. There are things you can do to promote this lifestyle and keep him active; you will greatly enhance his *joie de vivre* if you view things from his perspective.

Putting the zing back into the life of a bored cat

When I go and visit my clients I take with me my very old and battered consulting briefcase. It contains a wide range of rather tired but irresistible toys and is anointed with a thousand exciting cat smells. Without exception, the cats I visit make a beeline for the bag and spend ages sniffing, grabbing, dribbling, rolling and playing in response to its contents. All the owners, also without exception, say that they have never seen their cat so excited or animated. Isn't this a little telling? All I've done is brought new smells and novel toys into their homes. Admittedly it would be hard to reproduce the complex odour of my briefcase, but it wouldn't be hard to introduce some novel toys. The owners I go and see aren't bad owners (quite the contrary) but they just haven't thought about their cats' desire to explore new things or what really constitutes an appealing toy.

Games to play

Cats love to play; they are hardwired to catch prey, so anything that mimics stalking, pouncing and killing is going to be fun. There are two basic types of toys to promote play: interactive and solitary. The former you can wave in the air or trail on the ground, or move remotely with the aid of a few batteries, and the latter relies on your cat finding it suitably alluring to move it himself.

Interactive toys include:

- fishing-rod toys
- remote-control toys
- pieces of string/ribbon
- twigs and sticks
- laser lights (There is some controversy surrounding the use of laser dots and encouraging your cat to chase something so intensely bright and impossible to catch. Before turning it off I always end the game by shining the dot under the sofa, where the cat will find a small toy fur mouse that has been secreted there previously. Not all cats enjoy laser-chasing, so you should stop doing it if this is the case, but many will come running when they hear any sound that they associate with it, like a drawer opening or a key chain rattling.)

Solitary toys include:

- ping-pong balls

- real-fur toy mice
- any small object with catnip inside
- cardboard boxes
- paper bags
- anything your cat has found for himself and is playing with (e.g., plastic lid)
- screwed-up newspaper/paper/tissue paper
- corks

Many people spend a fortune on brightly coloured and imaginative cat toys only to leave them in a wicker basket in the corner of the living room, presuming that their cat will visit it and remove the toy he wishes to play with according to his mood. I certainly know some Orientals who do just that, but most cats have no concept of the basket being a toy box and the contents are meaningless to them. Most toys that are bought are never played with, so if you don't choose carefully, you are simply wasting money. If you know your cat well you have an idea of the sort of toy he prefers – for example, it has to be a particular size or texture or shape to appeal. Once you have established this you should then keep all the toys in a sealed container with a pinch of catnip in it, so that each toy can be brought out randomly to maintain its novelty. Your cat is far more likely to explore a toy under these circumstances than if it has been left on the carpet for days.

Challenging feeding

It's very easy over the years to allow the food to creep up in quantity and creep down in nutritional value in the name of spoiling your cat. If you suppose that small sachets of gourmet prawns with lobster jelly followed by leftover Chinese and a little double cream is a well-balanced diet, I beg to differ. Of course it's delicious, but indulging your cat's ever-increasing eclectic palate is not in his best interests. My Cornish cat Bink loves prawns; she probably would eat them exclusively if given the chance. She gets a nutritious and very palatable mix of good-quality dry and wet cat food that has everything in it, and the occasional treat. She has tried turning her nose up at the balanced stuff and hanging round the fridge, looking forlorn, but she soon gave that up when she realized that the prawn purveyor had a stronger resolve. This is one of the many occasions when it is absolutely essential to say no to your cat. See Chapter 4 if you want further bullying about feeding!

The quantity to feed is another factor, as it's important to adjust amounts according to your cat's lifestyle. If he goes outside in the summer and is generally more active, he needs more calories to maintain his body weight. If he is a couch potato who only seems to be up and active for about half an hour a day, then he has a greatly reduced calorific requirement (despite what he might tell you). Another common mistake is to presume that all cat biscuits are the same. Energy-dense kibble is packed with nutrients and every individual biscuit goes a long way, whereas other types of biscuit have a lower digestibility. Always check the feeding instructions and start at the lower end first – don't presume your cat has a fast metabolism. You can then adjust the amounts further if he seems to be gaining or losing weight.

I've spoken many times before about food foraging, and some entertaining problem-solving in this regard would certainly perk up the bored cat, but there are a couple of extra things that you can try that involve food. If you are feeding titbits (in moderation, I hope) you could try clicker-training (see Chapter 8) to train your cat to do something energetic, but essentially cat-like, to earn the favourite treat. I'm not a particular advocate of training your cat to perform, because I feel this sits uneasily with the whole concept of the cat as an independent and self-reliant species. However, I don't see any harm in getting your cat to jump up or climb something in order to get a treat. He would certainly have to work a lot harder to get something not nearly so delicious if he were hunting for a living.

You could also consider feeding your cat on a raised platform; walking into the kitchen and sticking his head in a bowl is not exactly the most exciting thing he has done all day. If he has to climb to get his biscuits or extract them from a receptacle using his paw or knock something over to spill the contents out, I think he may well enjoy the challenge. It's always best to leave some food in the usual place and use a portion of his daily ration for his experimental foraging. There will always be cats that resist all attempts at change when it comes to feeding and give withering looks to the person who dares to deviate from the normal routine. I would ignore the look of recrimination or the big sad eyes (unless of course your cat is one of the few that become genuinely distressed by this) and rest assured that, when he is hungry enough, he will find that extra food with a little bit of effort.

Hunting, shooting, fishing

Have you noticed that your cat hasn't been going out much recently? Did he always use to be out all day, mooching around and generally being a cat? Cats tend to reduce the time spent outside as they get older, but this is something more associated with old age than middle age. If a cat gets out of the habit of spending time outdoors or just feels safer inside, he is missing out on a great deal of fun activity. There are a number of factors that may contribute to this – for example, is he being bullied in his territory? Cats are usually very good at maintaining a comfortable distance between one another when they share an area, but occasionally one individual may decide to ignore the etiquette and actively seek trouble. Your cat may even give off vibes that indicate he is easily victimized and others may pick on him just because they can. I don't always recommend we fight our cats' battles, but in this case you may be able to restore some of his faith in the great outdoors. You could try:

- spending more time in the garden with him
- making a point of going out every day (weather permitting) and playing with him outside
- not feeding the birds or leaving any food around that could encourage other cats into the garden
- deterring other cats when they are seen in the garden (shouting, clapping, water pistol, for example)
- securing the garden (see Chapter 6)
- planting more shrubs and introducing pot plants to give him more camouflage
- giving him a tall perch outside. A shed or similar structure with an accessible roof may allow him to survey the territory and prevent surprise attacks. Be warned, though: if he likes it, the neighbourhood cats will like it too!

There are other changes that may have taken place outdoors that are equally off-putting. Your neighbour may have landscaped his garden, taking away that patch of weeds and tall grass that your cat loved so much. One of your neighbours may have a new dog that is preventing your cat from travelling along his regular thoroughfares. Any number of incidents may have unsettled him and he just needs a little encouragement and support to restore his confidence. Tune in to your cat and see if this could account for his middle-aged spread and general lack of zest.

Does my bum look big in this?

It is a sad fact of life that middle-aged spread affects our cats too. Obesity is one of the main health issues in cats between six and ten years of age. Many modern diets contain large quantities of carbohydrate and this converts to glucose, which is readily stored as fat if the calories taken in exceed the cat's requirements. Neutering cats, a necessary procedure nonetheless, slows the metabolism and results in fewer calories being necessary to maintain body condition.

Approximately 25 per cent of pet cats in the UK are overweight or obese. Extra weight puts a strain on the body, leading to diabetes, liver and pancreatic disease, cardiovascular and joint disease and even complications during surgery and anaesthesia, so it's not something we should view lightly.

Like so many health problems in life, prevention is better than cure. When your cat visits the vet every year for his annual check-up, he will be weighed and the vet will often assess his 'body condition score'. Weight in itself is not the best indicator for an ideal physique. Some cats have small or large frames and this will dictate whether they are naturally heavier or lighter than average. A body score describes condition that ranges from emaciated to grossly obese and covers all variations in between. Ideally the cat should have a well-proportioned body, with an obvious waist behind his ribs. You should be able to feel your cat's ribs, but not see them, as you run your hands along his sides. If this isn't the case, then the obvious signs of obesity are:

- The 'coffee table' look: when observed from above a flattened appearance over the back that indicates the presence of large pockets of fat either side of the spine. If you can (hypothetically) balance a cup of coffee on your

cat's back, he is definitely obese (please don't try this at home).
- Dirty or unkempt bottom, tail or hind legs. Cats with too much body fat are not as flexible as they should be and many find it hard to reach their nether regions to wash efficiently.
- Excessive dandruff. Obese cats often have dandruff in their coats.
- Loss of agility/mobility. Obese cats don't jump to great heights and are reluctant to exercise.

Some cats do have an obvious apron of skin that hangs from their belly; my Mangus has a rather pendulous one that the nurses at our vet practice euphemistically refer to as her 'water features'. This isn't necessarily an indicator of obesity, as in Mangus's breed, and some others, this can be genetic (honest!) and other cats will have it after pregnancy or losing weight.

Studies show that there are cats at higher risk of obesity than others. Statistically the most susceptible group are non-pedigree neutered males that live indoors, but any cat that eats more than its body needs will become obese.

'Weight Watchers' for cats

You don't have to wait until your cat is waddling before you decide to put him on a diet. If his waist is going or he's put on weight since last year and he's become a little lazy, now is the time to start adjusting his diet before the situation becomes seriously threatening to his health. Light varieties are available of many foods that have all the nutrients your cat needs but fewer calories. Your vet may advise you to switch to one of these light diets or merely to reduce the amount of your cat's regular food. Introducing all the feeding techniques and activities that I've suggested before and strictly monitoring your cat's intake will soon get him back on track.

If, however, your cat is approaching coffee-table proportions, you may need to take more drastic measures. Your vet, or a nurse in the practice who is trained in obesity management, will supervise any weight-reduction programme. Your cat will be weighed accurately and a body score will be given to show the starting point. If your cat's ideal weight is 4kg, for example, and he now weighs 5kg, you may think 1kg is not much of a problem. If, though, you see it more as 25 per cent excess body weight (that's more than two stone for the average person), you can probably relate to what your cat has been carrying round on his back for a while. The nurse or vet will then set your cat a target weight (this won't always be his ideal weight straight away if he is very obese) and a date to aim for, several months hence, to ensure that the loss is gradual. If your cat loses weight too rapidly, muscle tissue will be lost with the fat. A successful weight-loss programme achieves a loss of between 1 and 2 per cent of the total body weight per week as a maximum.

Veterinary practices all have balanced proprietary 'prescription' diet foods available for the management and reduction of obesity. One of the options is a food that contains high levels of fibre, specifically the type that ferments slowly. This will increase the bulk of the diet and promote a feeling of fullness without adding calories. You may find that the quantity of food looks larger than normal but a lot of this high-fibre diet will pass straight through your cat. It's certainly worth noting that his stool production will be significantly greater than usual.

Some cats find the increased fibre unpalatable and it can cause bowel disturbances, but there are other weight-loss diets that are lower in fat and calories. Either method, given in the right quantity and as part of a weight-reduction programme, will produce good sustainable weight loss; your vet will be best able to pick the one that's right for your cat.

Unfortunately, just feeding the right food in the right quantity is only part of the solution to obesity. Many owners are dissatisfied with their cat's weight loss but, once questioned, soon realize that there are other considerations. The most effective weight-reduction programme is a holistic process, with genuine lifestyle changes for your cat.

THE GUIDE TO SUCCESSFUL WEIGHT LOSS

- Feed a diet that contains reduced fat and calories but all the necessary nutrients. Your vet should recommend the appropriate food.
- Weight-reduction diets are available in wet and dry forms, so use either or a combination of both – whatever your cat would most enjoy.
- Your cat's weight loss should be monitored by your veterinary practice, always ensuring he is weighed on the same scales to give accurate measurements.
- All titbits and table scraps should be excluded from his diet; these usually contain high calories.
- The recommended daily amount to feed should be carefully weighed and no extra should be given no matter how insistent your cat is.
- If you can't resist giving your cat a treat, then use a few biscuits from his daily ration for the purpose.
- The food should be divided into portions and small amounts given throughout the day in problem-solving receptacles to allow your cat to 'hunt' for his food.
- Single kibble can be rolled across the floor, or even thrown up or down stairs, for him to 'catch' and eat.

- Food can be placed on high surfaces to encourage him to climb.
- Provide exercise, starting at five minutes twice a day, by encouraging your cat to chase fishing-rod toys, small balls or string, for example.
- If your cat is reluctant to play, then bumping feathers at the end of a fishing-rod toy into his body will inevitably get a response!
- Offer him a pinch of catnip on the floor prior to playtime; this excites some cats and might make him keener to be active.
- Tuck a long ribbon into the back of your clothes and let it drag along the floor as you move around the house; this may do it!
- Don't give up. Your cat will eventually start to play and he will then enjoy it.

If your cat is overweight you need to monitor or change his diet; if you ignore it, his middle-aged spread will become obesity with time. No matter how cute you think pictures of barrel-shaped cats look and how you marvel at tabloid headlines such as 'Is this Britain's fattest cat?' please don't ever think this is a good thing. You may have thought, 'Oh, he loves his food!' but this is really missing the point. These fat cats will probably suffer disease or even die prematurely purely because they are obese. When a cat is that fat it will continue to eat because that's all it can do; it can't walk, it can't groom and it can't play. Surely you don't want that for your cat?

Special needs of the elderly

No one likes to dwell on the effects of ageing, but we ought to focus on it for a few moments so that we understand what's happening to our cats (and to us!) as the years pass. Your cat's fur starts to lose its shine, the skin becomes less elastic and a few white hairs appear. Hearing, sense of smell and sight deteriorate and eventually your cat's memory may be affected. Sleep increases and activity decreases as the muscles and bones weaken and the immune system becomes less effective.

As the body grows older your cat's behaviour will also change as he adjusts to the deterioration in his ability to hunt, fight, climb, jump and generally defend his territory. In the Elderly Cat Survey, in which owners of 1,236 cats over twelve years of age took part, general trends in the behaviour of older cats were noted.

- They go outdoors less (reported by 55% of surveyed owners).
- They hunt less (47% had stopped completely).
- They sleep more (40% slept eighteen hours plus and 57% slept for up to eighteen hours a day).
- They vocalize more (66% of owners reported increased sounds to get food, attention, etc.).
- They become more sociable and affectionate (81% of owners reported this).
- Grooming becomes less frequent and less effective (24% groomed only occasionally or not at all).
- They play less (63% of owners reported that their cats played only occasionally or not at all).
- Their toilet habits change (29% of the cats had toilet 'accidents' in the house).

If you understand the likely physiological and emotional changes that take place in your cat you should be well equipped to deal with them. It's important to remember that old age is not a disease and that with a few little adjustments you are able to make his twilight years comfortable and enjoyable.

Beds and 'Zzzzeds'

As a favourite pastime of the elderly is sleep, you need to ensure that appropriate provisions are in place to make it a pleasant experience. The majority of owners in the Elderly Cat Survey indicated that their cats had favourite places for their forty winks. The top ten preferred locations were all the most comfortable furniture in the house, including the master bed, and just about anywhere else in the sun or near a source of heat. This isn't surprising as older cats tend to have less fat in their body and their ability to regulate their temperature is often greatly reduced. Even when you feel the house is at a comfortable temperature, your elderly cat may still be in need of some extra warmth.

If you don't relish the cost of leaving your heating on all day (who would?), you could invest in an electric heated pad or one that can be heated in the microwave. This can be placed on a soft chair, within a cat bed or inside a padded cardboard box with the front section cut away for ease of access. If this is placed in a quiet draught-free area, where your cat can be safe and sleep undisturbed, you will find he will take to it, luxuriating with a look of blissful contentment on his face. If you don't fancy that, then ensure you make good use of the airing cupboard, underfloor pipes, Aga or Rayburn (if you are fortunate enough to have one), boiler or any other household appliance that produces a continuous source of heat.

When you are looking to create the perfect bed for your cat it's important to remember a few other special considerations:

- Make sure the bed is thick with fleeces, blankets or similar

material, ideally with thermal qualities. This will prevent any sores developing over any prominent bones your cat may have.
- Some older cats find curling up a little tricky, particularly if they are arthritic and stiff. Provide a bed that is big enough for him to stretch out to give him added comfort.
- Most cats like resting in elevated places but it can be difficult to get there if they can't jump. Check out all your cat's favourite sleeping places and make sure there are gradual steps up to it, using furniture such as footstools and small tables.
- If he favours a radiator hammock, you may need to provide steps up to this too.
- Don't disturb your elderly cat when he is sleeping; older cats often sleep lightly and they are easily woken.
- Don't be surprised if your cat changes his preferred resting place from time to time; this is perfectly normal and should only raise warning signs if he is seeking cold places, as this can be a sign of disease, such as stomach tumours or hyperthyroidism. A trip to the vet's will put your mind at rest.

'Use it or lose it!'

Your cat may be hunting and exploring less but that's no reason for him to give up the pleasures of physical and mental exercise. Older cats may be a little stiff and less agile than they used to be but regular activity will help to retain muscle mass and aid circulation. This is even more important if your cat has turned to food for entertainment and is starting to gain weight.

I am not suggesting for one moment that you encourage your cat to leap about like a mad thing and get hopelessly out of breath. This will either result in you doing lots and your cat doing nothing, apart from staring at you in utter disbelief, or see your poor cat off once and for all. Gentle and regular playtime for short periods is the way forward for the elderly. You may get exactly the same charging around that he used to do, but now it's for significantly shorter periods, or he may lie on his back and wave his legs in the air to catch the feather you are dangling above him. Either way, it's exercise.

For those cats exhibiting signs of senility (see below), environmental stimulation in the form of hunting and foraging games and structured interaction can lead to an increase in cognitive function and a better quality of life.

Exercise and the disabled cat

Not all disability sufferers are elderly but problems are more likely to occur in later years. Amputees, in my experience, cope admirably with their disability, so it's almost not worth mentioning them as cats with special needs. If, however, the amputation takes place later in life it may be more debilitating for the patient and the sort of gentle exercise recommended for the geriatric cat would probably be appropriate.

The most common disabilities of the elderly are undoubtedly deafness and blindness. Hearing loss tends to be gradual and the owner usually misses the signs until the cat is stone deaf. In this case, play must be initiated visually, but deafness shouldn't hinder their ability to chase and catch.

Unfortunately blindness can have a sudden onset and be extremely distressing for the cat. If, though, vision is deteriorating gradually, a trial-and-error technique to discover which games he still enjoys playing is probably your best bet.

Try not to over-compensate, as most cats adjust well to most disabilities. All the same, it's probably wise to curtail your cat's activity outside the relative safety of your own garden.

Routine, routine, routine . . .

One element of feline old age that is common to most is an increasing insecurity and dependence. Cats are born survivors but they soon know when everything isn't working quite so well and will turn to their owners for comfort. The great gift you can give them at this time is routine. By maintaining the furniture layout of the rooms and having set times to eat, play and cuddle, you will give your cat a strong sense of security at a time when lack of familiarity means trouble. Old cats don't handle stress quite so well as the youngsters, so keeping everything just the way they like it is the best way to combat this potential problem. Often holidays are an issue when you have an elderly cat, but providing a cat-sitter – ideally a friend or member of the family who can live in – to care for him in your absence should be sufficient to ensure he's fit and well when you get back. For most cats, suddenly having to visit a boarding cattery in their senior years is a recipe for disaster, no matter how luxurious the establishment.

The daily constitutional

Most old cats with access outdoors still like to patrol in clement weather, despite the fact they are spending less time outside generally. If your cat has previously used a cat flap you may need to review its suitability for his advancing years. Some find it difficult to negotiate a flap, so you may have to play doorman from now on. It's probably wise to make sure he has a microchip or a secure collar for those senior moments when he wanders off and can't remember how to get back.

New additions

Older cats not only hate change, they also usually resent

newcomers. Many owners think a new kitten will give their ageing moggie a new lease of life but this is rarely the case. Think long and hard before you decide to inflict a kitten on your oldie; he may not be as grateful as you imagine and he could become genuinely distressed.

Toilet trouble

The confidence to toilet outside in the presence of strange cats disappears with age and many cats, having always used the garden, suddenly have 'accidents' indoors. Old cats are also reluctant to relieve themselves outside in cold, damp or windy weather, so an inside toilet isn't much to ask, is it? I always recommend that an indoor facility be provided for the elderly, ideally before the poor soul starts to poo behind the television because he is too frightened to go outside.

Diet and the elderly cat

Of the owners taking part in the Elderly Cat Survey 56 per cent said that their cat's appetite had stayed the same in old age; 25 per cent said it had increased and 24 per cent that it had decreased. Periodontal disease affects 85 per cent of all cats to one extent or another and this can dramatically affect the appetite if the problem remains unresolved for any length of time.

If your cat is fit and healthy there is probably no medical reason to change his diet. The antioxidants and polyunsaturated oils added to some senior diets may be beneficial, but there is some debate about this. Some evidence in other species suggests that antioxidants such as vitamins A (beta-carotene), E and C may play a role in protecting against some normal ageing processes, so this may also apply to cats. Some studies have shown that senior cats do not digest and absorb fat as well as younger cats, so there may be a need to introduce more digestible fat to get the same amount of energy. There used to be a tendency to restrict protein intake in older cats, just because they were old, but this isn't advisable. Cats need more protein than do many other animals; inadequate amounts can impair the function of their immune system. Unless the cat has a health condition that requires protein restriction, he should not be given a protein-restricted diet. If your cat is approaching his senior years with no apparent changes in his demeanour or appearance, it's probably advisable to ask your vet whether there's any need for a dietary change. When I have discussed this with my vets in the past they have always recommended diets appropriate for my healthy cats' life stage, but check with yours before making that decision.

Various disease processes may, however, require dietary changes to lessen the effects of the disease or slow its progress. There are a number of 'prescription' diets on the market specifically intended for cats with complaints in old age – for example, constipation, diabetes mellitus, colitis and heart disease. If your vet has diagnosed your cat with inflammatory bowel disease, or even colitis that is not responsive to a high-fibre diet, he or she may recommend a diet that has highly digestible sources of protein, fat and carbohydrates. If your cat has chronic kidney failure, then your vet will recommend a diet with highly digestible protein, so that the kidneys have to work less hard to excrete the waste products resulting from its metabolism. There are even diets now that are specifically formulated for cancer sufferers, with increased Omega-3 and beta-carotene. Dental disease is also helped by certain dietary products with additional enzymes and a particular kibble structure that scrapes against the teeth to aid in the removal of plaque. However, once your cat reaches a certain stage and teeth have been extracted, he may need a soft tinned diet instead.

Unfortunately, your providing all these fancy diets doesn't mean that your cat will eat them. It's so frustrating when your vet has convinced you that your cat will live longer and feel better if he eats a 'kidney' diet (for example) but you find it virtually impossible to convince the cat! I think the problem here is multi-factorial:

- Older cats' sense of smell deteriorates with age, so they tend to be attracted to foods with a stronger smell to stimulate their appetite; prescription diets are not always the yummiest.
- New flavours or textures are often rejected; old cats love the routine and security of familiar things.
- Cats manipulate their owners well by the time they get to this great age; they know you have prawns in the fridge and they know you will offer them if they reject their cat food. (Half of the owners in the Elderly Cat Survey had been 'trained' to feed their cat on demand.)
- Owners are understandably concerned about their cats' welfare and keen to get them to eat the life-saving food; the sense of urgency transmitted is guaranteed to make the cats believe it is poison.

All this can be a source of much distress for both owner and cat. There are things you can try before you give up:

- Choose the consistency that you know your cat likes – for example, if the prescription diet comes in a tinned variety, your cat may prefer that to biscuit.
- Mix the new food gradually with your cat's previous diet;

this will allow his palate and his digestive system to adjust slowly.

- Feed your cat little and often and walk away when he is eating; that way he won't feel any pressure.
- If you are blessed with a cat put on this earth to please you it may be helpful to stay with him when he is eating and give him the odd appreciative stroke to show how grateful you are that he's eating his 'special medicine' (this won't work with most cats).
- Consider heating up the wet version of a prescription diet to body temperature to release more of the flavour aromas.
- If your cat flatly refuses to eat it you could ask your vet if there are any supplements or home-made diets that may provide your cat with the necessary nutrients.

If you fail, no matter what technique you employ, please don't be disheartened. If your cat is poorly there is little point in upsetting him further by attempting to force him to eat something he clearly doesn't like. It is probably better to make his life as comfortable as possible for as long as he's got, rather than extend a rather miserable last few months or years. I am a great believer in feeding the very sick cat whatever he wants to eat at the end.

Don't forget water

Older cats tend not to drink sufficient quantities of water and this can be a major problem if your cat suffers from constipation or kidney disease. You may be able to encourage him to drink more by providing additional sources of water away from his feeding area and even adding a little flavour – for example, a spoonful of fish stock – to encourage fluid intake. You could even try adding lukewarm water to his wet food.

Getting the elderly cat to eat more

It can be a problem getting your older cat to eat sufficient food, particularly if he is suffering from ill health. You could try:

- heating the wet food as described above; take care to stir the food afterwards and test the temperature first if you are using a microwave
- adding water (not brine) from tinned tuna to increase the flavour of the food
- feeding your cat 'little and often' and making mealtimes an opportunity for petting him if your cat has become more affectionate in his old age
- switching from dry to wet food to see if your cat finds this more appealing. You could speak to your vet about giving your cat a high-calorie food that is rich in nutrients, so that a little goes a long way. These foods are traditionally fed to convalescing animals
- feeding your cat away from children, other younger cats, the dog or anything else that could make him feel less than comfortable
- sprinkling a tiny pinch of dry catnip on his food – often this stimulates appetite

Preventative care and general maintenance

Forewarned is forearmed, as they say, so regular examination of your older cat will detect changes that may need to be tackled straight away. It's also the perfect time to carry out 'essential repairs':

- Check your cat's nails weekly. Elderly cats are less able to retract their claws, which get caught in furniture and carpets. They can also overgrow and stick into their pads.

Regular trimming will be necessary, either at home (see Chapter 5) or at the vet's surgery.

- Older cats are less able to groom efficiently, so you may need to wipe away any discharge around the eyes, nose or anus using cotton wool moistened in warm water. You will also probably need to brush your cat using a soft brush and fine comb, but be sure you are gentle as he doesn't have much padding and vigorous combing can be painful. At this time you can also check for lumps, bumps, sores or anything else that merits attention from the vet.
- If your cat is longhaired he may be grateful for a 'short back and sides' around his bottom and 'trouser' area to make washing easier.
- Old teeth and gums can be a real problem, so it is wise to take a look, if your cat allows it, to check for any growths, reddening of the gums or evidence of dental disease; raging halitosis is a sign that all is not right.
- Check the indoor litter tray for blood in the urine or stools, consistency of stools or other indicators of disease. This is probably not the most glamorous thing you'll do all day but it's an important source of information about your cat's general health.

There are a number of general warning signs in the elderly that merit attention and a visit to the vet, namely:

- loss of appetite
- weight loss
- excessive drinking
- stiffness, lameness or difficulty in jumping up
- lethargy
- sudden appearance of lumps or bumps
- balance problems

- toilet accidents
- drooling
- difficulty passing urine or faeces
- disorientation or distress

Many vets still advise regular booster vaccinations for old cats, although progressive practitioners extend revaccination intervals and assess the needs of the individual rather than following blanket rules. This doesn't mean that annual visits and check-ups are a thing of the past; preventative health examinations are particularly important for the elderly and your vet may even recommend that they take place every six months rather than annually when your cat gets to a certain age.

Diseases of the elderly

The signs of disease don't always manifest themselves as the obvious causes for concern listed above; many early indicators are in subtle changes in the patterns of your cat's behaviour.

Diseases that are commonly seen in the elderly cat include:

- hyperthyroidism
- renal insufficiency (chronic renal failure)
- arthritis
- dental disease
- oral tumours
- diabetes mellitus
- inflammatory bowel disease
- FIV infection
- systemic hypertension
- liver disease
- urinary tract infection
- cognitive dysfunction syndrome (senility)

There are other issues that can also affect your elderly cat's behaviour – for example, bullying by other cats, the general insecurities of old age, impaired hearing and vision, to name but a few. The owners completing the Elderly Cat Survey were asked what conditions their cats suffered from as oldies and the top five were listed as: 1 arthritis; 2 chronic renal failure; 3 deafness; 4 blindness; 5 hyperthyroidism.

Arthritis

It's probably worth focusing for a while on the condition that owners believe is the number one problem but has traditionally been under-diagnosed (or its significance vastly under-rated) in veterinary practice. In a recent study of ninety-nine cats over the age of twelve, 65 per cent were found to have radiographic evidence of osteoarthritis (OA). Vets have previously considered that cats don't suffer from OA like dogs, mainly because their small frame means there isn't so much strain put on joints and they don't show obvious lameness or even obvious signs of pain.

Research is now showing that they do indeed 'suffer' from OA, but they exhibit that suffering in a different way. This is a subject particularly dear to my heart, as my lovely Annie had severe OA for some years and it took a while for the vet to appreciate the significance of it. As a result of that experience, I believe it's really important that you appreciate it may be an issue for your cat too.

The signs of OA that you should look out for are:

- a reduced ability to jump or climb
- 'bunny-hopping' downstairs or difficulty climbing stairs
- a 'tucked-up' gait with the spine arched
- stiffness and a shuffling or stilted gait
- lameness
- reduced activity or social interaction

- joint swelling or pain or reluctance to be handled in a particular area
- reduced range of movements in the limbs, spine or neck
- reduced or incomplete grooming
- inappropriate urination or defecation

A combination of a number of these factors would be enough for me to take my cat to the vet for an X-ray.

The best way to judge the impact of OA on cats is to look at the effect of painkillers; studies show that, once cats are receiving medication, there is a significant change in their behaviour. They leap to high places, are no longer stiff or lame and are generally much more active.

If your cat is diagnosed with OA, then the treatment options will include some or all of the following:

- non-steroidal anti-inflammatory drugs (NSAIDs)
- glucosamine and chondroitin supplements
- weight reduction in overweight or obese patients
- environmental changes – steps or ramps up to high places, daily grooming by the owner, low entrances to litter trays, removal of cat flaps in favour of opening doors, etc.
- a diet specifically designed to assist in the management of OA (containing specific fatty acids proven to be beneficial in other species)

Senility

As cats are living so much longer these days conditions have arisen that would never have been seen by previous generations. We now have an ageing population of cats and some suffer from senile changes similar to those seen in human patients afflicted with Alzheimer's disease. The condition, referred to as cognitive dysfunction syndrome (CDS), is diagnosed by your vet based on symptoms and the ruling out of other potential causes.

The various behavioural changes that may indicate a diagnosis of CDS include:

- a change in sleep/wake cycle
- excessive vocalization, particularly at night
- decreased appetite
- disorientation or confusion – e.g., getting lost outdoors
- altered memory – e.g., failure to use litter tray
- change in social behaviour – e.g., increased dependency or aggression
- decreased response to stimuli in the environment
- reduction in grooming
- staring into corners
- repetitive pacing

Once obvious and severe signs of senility are seen it is difficult to prevent further deterioration. Environmental changes should be kept to a minimum, as senile cats cope poorly with change. Furniture, food bowls, water, beds and litter trays should remain in familiar locations, but in some cases cats will actually be reassured if all their relevant resources are contained within one room. I have counselled many owners whose senile cats have kept them awake with nocturnal wandering and howling. These cats have become significantly less stressed and generally happier when a core area has been created within one room and they are shut into it at night.

Various drugs and dietary supplements have been used experimentally to assist sufferers, so your vet may recommend a special diet, supplement or medication based on your cat's symptoms.

Quality of life

It's impossible to talk about cats getting older without addressing their quality of life towards the end. Most cats these days do not die a 'natural death'; this isn't a bad thing as it can be distressing and painful and not something I would wish to put my own cats through. The whole idea of dying peacefully at home is seldom the reality.

It is, however, notoriously difficult to assess pain and discomfort in cats and, as with conditions such as osteoarthritis, their wellbeing is often better assessed by looking at changes in their normal patterns of behaviour.

In previous books I have written about the use of 'Activity Budgets' to monitor the time your cat spends performing normal behaviour such as sleeping, resting, grooming, eating, playing and socializing. As cats are creatures of habit, these will tend to follow a familiar pattern, with slight changes associated with, for example, bad weather, visitors or other general disturbances. These patterns change, often significantly, when cats are in pain. Certain changes in character or posture that may indicate suffering can also be noted. These include:

- sudden-onset fear response
- aggression
- reluctance to be handled
- furrowed brow
- hanging head
- 'glazed expression'
- sternal recumbency (resting upright on the chest)
- 'tucked-up' abdomen
- arched spine
- lameness, stiffness
- rubbing painful areas
- head-pressing

- inappropriate elimination (soiling in the house)
- localized over-grooming
- increased/decreased vocalization

If your cat is exhibiting a number of these character and postural changes and his pattern of behaviours shows less activity, poor appetite, failure to groom, toilet accidents and a general withdrawal from social contact, for example, then he may be suffering. 'Quality of life' is a balance between good things and bad things, pleasant and unpleasant. Providing the balance swings in favour of the pleasant things, it can be said that there is a quality of life. If the only 'pleasant' thing becomes life itself, then that is no justification to carry on. It's an extremely tough decision for you to make, but if your cat is terminally ill it may be necessary, sooner rather than later.

The questions you need to ask yourself are:

- Is my cat's pain incurable?
- Can drugs alleviate his continual discomfort?
- Are there no further treatment options for his condition?
- Is he having toileting difficulties or having accidents?
- Is the administration of the medication causing him distress?

If you end up with the answers you don't want to hear then it's probably time to call it a day, for his sake.

Euthanasia

Your vet will guide and support you towards this final decision but it is, ultimately, one you have to make yourself. You have the option to wait one last day, if you feel that's appropriate, or organize things straight away. You will also have a choice of having your cat 'put to sleep' at home or in the vet's surgery.

Many cats hate the prospect of a trip to the vet's, so it may be kinder to ask your vet to visit you at home.

Euthanasia is basically an overdose of anaesthetic that is administered into a vein in your cat's foreleg. If he is very sick and his veins have collapsed, then your vet may inject the solution into his heart or kidney. If there is any risk that your cat will be fractious, he may be sedated first before the final injection is administered. Most cats, though, tolerate the process well and offer very little resistance. You may find the prospect of holding your cat while the procedure is carried out distressing; under these circumstances it is best to ask for a nurse to be present to assist the vet in your absence. Unfortunately, distressed owners disturb the cats and the end is often more peaceful if they aren't there.

Many owners ask if other cats within the household should see their dead companion. Twenty-eight per cent of the owners in the Elderly Cat Survey reported that their cats had shown visible reactions to the loss of another cat in the household. Some owners believe that cats do not mourn if they see the dead body; yet others report that their cats have been distressed, probably due to the strange smells from the veterinary practice. There is really no conclusive proof either way, so I think it best that owners do what feels right for them at the time.

Cats react to varying degrees when they lose a companion, but their response isn't always negative; some blossom after a period of time, relishing life without the departed cat. It is rarely as traumatic a loss for them as it is for their owners.

When the time comes . . .

You have a number of options when the time comes; your vet will gladly help you in making the decision and probably already have an arrangement with a local pet cemetery or crematorium who will look after your cat's remains with dignity. In my experience, establishments 'behind the scenes' take great care to treat the animals as every owner would want them treated. There is an organization called the Association of Private Pet Cemeteries and Crematoria that is currently the only one in the UK that sets specific standards for cremation and burial of pets. This being the case, I am sure all reputable companies will show allegiance to a scheme that promotes best practice.

The most common options are:

- Burial at home – you can bury your cat in your garden unless local by-laws say otherwise. The grave must be at least three feet deep to deter scavengers.
- Individual burial – your cat will be buried within a distinct plot in a pet cemetery that you can visit.
- Communal burial – your cat will be buried with no distinct plot, although this service is relatively uncommon.
- Communal cremation – your cat will be cremated with several others with no separation; the ashes are then buried or scattered in an area specifically intended for that purpose.
- Communal cremation with separation of ashes – your cat will be cremated with several others but there will be some separation. The ashes are removed one at a time but some mixing of ashes may occur. The ashes are then returned but this doesn't absolutely guarantee you have your pet's ashes and no other.
- Individual cremation – your cat is cremated singly and all ashes are returned to you or, on request, scattered in an area specifically intended for that purpose.

If you are not satisfied with the more conventional choice of burial or cremation for your pet, then there are a startling variety of options for the more adventurous (or, dare I say, eccentric) bereaved cat owner. Most of these alternative 'memorials' are, admittedly, embraced more completely in America, so your cat may have to make one last, transatlantic trip before his final rest.

I have always said, joking apart, that how we remember our beloved pets is a very personal matter. It's not for me to judge, because something appearing to be in appalling taste to one person may be a comforting and beautiful reminder of the dear departed to another. Whatever your own moral and ethical sensibilities, somebody somewhere thinks all of these services are a great idea.

Here is a brief overview of the alternative options.

Freeze-drying

Yes, you can preserve your cat for ever by using modern freeze-dry technology! The service is currently only available in the USA and I somehow doubt that it's coming to the UK any time soon, but it's a fascinating process. The companies offering this service describe it as the ultimate way to ensure owners 'never have to let go' of their dead pets (is that healthy?). The body is posed, supported by a special framework and placed in a sealed chamber, where, over a period of ten to twelve weeks, all the moisture within the tissues is frozen and sucked out. The owner is requested to send photographs so that the technicians can place the cat in a pose that looks natural.

The somewhat lighter-than-normal-freakishly-stiff body that was once your cat is then permanently preserved and, providing you don't leave it outside, spill something on it or generally mistreat it, the companies claim that it will stay that way long after you are gone.

I have a number of personal reservations, including one that was highlighted by a satisfied customer who placed feedback on one of the freeze-dry companies' websites. The lady wrote to say how delighted she was with the frozen remains of her much loved nineteen-year-old cat. She sent photographs to show her other two cats investigating their frozen dead companion. She remarked how amusing it was to see the reaction of the cats, as they had hated the older cat when he was alive and were appalled that he was back. The pictures showed two very disturbed live cats that will probably never be the same again. Such is the nature of love; it comes in many different guises.

Cryopreservation

Now, this is serious science, or is it? An American company offers to suspend your cat in liquid nitrogen, after various processes have been carried out to remove the blood and guard against damage from freezing, so that your cat can be brought back to life at some stage in the future. The flaw in this plan is that it cannot work with any creature that is truly brain dead. The whole concept is based on the belief that science will advance to such an extent that it will enable the cryopreserved animal to be brought back to life. No mammal has yet been preserved in this way and subsequently revived, although dogs and monkeys have had their blood replaced with a solution, been cooled to below 0°C and then returned to life with no obvious ill effects. Rabbit kidneys have been completely 'vitrified' to −135°C, re-warmed and transplanted viably into a live rabbit, but this isn't really proof that Sooty could one day live again.

Cryopreservation is an expensive process that hangs on the premise that we as humans have the power to defy death. I'm really not sure it should ever be thus.

Cloning

So far six cats, to my knowledge, have been cloned for rich pet owners. The company in question, based in America, has reassured owners that they can reproduce an animal that is genetically identical to their lost pet. The fee for this service, at the time of writing, stands at $32,000. Before you start saving madly, however, I would point out that the cat may be a carbon copy genetically but your old cat is a one-off and if you feel you will be able to live your life together all over again you will be terribly disappointed. Over the years I have seen many cases of owners complaining about their cats' vague behavioural problems that require my assistance. During the consultation I have established that the cats were chosen in the first place because they uncannily resembled previous much loved cats that have sadly died. The biggest problem in these cases is that the 'replacement' fails miserably to live up to the owner's expectations and the relationship suffers as a result. I have an uneasy feeling that this would be the case with a cloned animal too.

Diamond memorials

If you want to blow a lot of money on a memorial that is as personal as cloning, not furry but sparkly, you may wish to turn your cat into a diamond. The technology exists to create a gem from the carbon of your cat's cremated remains.

Taxidermy

This now seems like a conventional alternative given the sci-fi qualities of cryonics, cloning, diamonds and freeze-drying. About twenty-five years ago I met a taxidermist; he invited me to see his workshop as he had been commissioned to 'stuff' (or should I say 'mount') a number of tigers after they passed away in various zoos. His workshop was amazing – full of gazelles, antelopes and big cats – and I remember asking the question 'Do you do pets?' He replied that he certainly didn't and never would, as it was virtually impossible to reproduce the posture and expression of an animal that was a family member rather than a wildlife exhibit. This is a common feeling among many taxidermists, although there are a few in the UK who do offer the service to pet owners. Unfortunately, with the best will in the world, even a skilled taxidermist cannot make an old or sick cat look particularly well again.

A taxidermist in America offered a slightly different alternative: to turn your departed cat's treated pelt into a 'soft, huggable' cushion. This lady, after posting details of this service on the Internet, has since gone into hiding, owing to death threats from animal rights activists who clearly found her suggestion in the worst taste. Not everyone understands the need of some individuals to mourn the death of their pet in their own unique and personal way.

Portraits and mementoes

I have personally chosen to remember my cats with photographs and oil paintings. Many excellent portrait artists can be commissioned to paint a picture of your cat, using any photographs you are able to supply. Fur can be clipped from your cat and kept in a locket with a small photo as a permanent reminder.

Coping with bereavement

Research shows that up to 75 per cent of owners experience difficulties after their pets die. In my experience this is undoubtedly the case, with many exhibiting signs of grief that they admit they never showed so openly for human loved ones. I really cannot believe this is an indication that we love our pets more than our family, but I do think it may just be that in this country we find open displays of grief more acceptable for pet loss than family funerals.

Many bereaved owners are not fortunate enough to have supportive people around them and find themselves in a position of having to cope with their grief in complete isolation. This is an incredibly difficult task for anyone, no matter how strong, and I would always recommend reaching out for a helping hand. Modern veterinary practices and your doctor's surgery should be able to provide you with details of local or national bereavement support groups; all you have to do is ask.

Sometimes it's enough to know that the strange physical and emotional symptoms you are experiencing are normal and that millions of other people have been there before and come through unscathed. It can be frightening when you honestly believe you are suffering a unique experience.

The expression of grief

Grief being such a very personal thing, any number of the following symptoms can be experienced as part of the perfectly normal but painful journey to recovery.

Physical signs

- shock
- continual crying and a permanent lump in the throat
- shortness of breath, tightness in the chest
- nausea
- loss of appetite or increased appetite
- exhaustion, aches and pains

- dizziness
- insomnia or disturbed sleep

Emotional signs

- sadness, depression
- anger, irritability
- anxiety
- relief
- loneliness
- helplessness

Intellectual signs

- confusion
- inability to concentrate
- visual and auditory hallucinations (e.g., cat crying)
- need to talk about and rationalize the loss
- preoccupation with death and the afterlife

Social signs

- withdrawal from contact with others
- rejection of help from others

There is no structure or set pattern to the journey through grief; from a personal perspective I can honestly say I have experienced all these signs to one degree or another when I have lost a cat. Some people move backwards and forwards through various stages, or get stuck in the angry stage, for example, for an impossible length of time.

Whatever happens, the advice is to accept that grief is inevitable and to succumb to it. Problems can occur when you don't allow these feelings to express themselves fully. Eventually the sadness and the tears do go away as they are replaced with fond memories of all the happy times you shared with your cat.

The following are suggestions to help you cope with bereavement:

- Plan ahead and have an idea in your mind what arrangements you wish to make at the end – e.g., burial, cremation, return of ashes.
- Accept that grief is perfectly normal.
- Accept help from family and friends and don't take to heart anything that is said by someone who doesn't understand owner/pet relationships.
- If you are relatively isolated in your social life don't be afraid to talk about your feelings to staff at your veterinary practice, your priest or a professional bereavement counsellor.
- Make a scrapbook or have your favourite photograph of your pet enlarged and framed; it may help.
- Try to ensure you are looking after yourself while you are grieving; it won't help the pain go away if you stop eating.
- Allow yourself time to mourn but try to return to normal routines as soon as possible.
- Understand that recovering from grief does not mean you forget about your lost cat or you love him any less.
- Considering acquiring another cat is not disrespectful to the memory of the departed. However, don't expect your next to be the same – all cats are individuals!

CONCLUSION

THIS BOOK HAS BEEN RESEARCHED THOROUGHLY AND TO THE very best of my knowledge it gives a comprehensive account of the facts. I have been unable to resist the opportunity to pepper the contents with my own opinions too. Yours may differ from mine but often facts and opinions have a number of different interpretations; there are no right or wrong approaches to certain things. However, regarding some topics, there is an awful lot of nonsense still circulating about cats and their care, so I hope I have dispelled some of these myths in this book.

It was always going to be a dangerous gamble calling this book *The Complete Cat,* but I reckon I've gone a pretty long way towards providing all the practical, important stuff you will need on a day-to-day basis. Some extra information is to be found in my three previous books, *Cat Confidential, Cat Detective* and *Cat Counsellor,* so look there first if you can't find something here.

APPENDIX 1

What Does It Mean When My Cat . . . ?

UNDERSTANDING CAT BEHAVIOUR

ALTHOUGH EVERY CAT IS AN INDIVIDUAL THERE ARE occasionally times when cats behave in a similar way for a particular reason. There are also behaviours exhibited by our beloved cats that are hard to interpret and often puzzle us. Here are a few of the most popular questions posed by cat lovers and a quick insight into normal cat behaviour.

Why does my cat rub round my legs?

Your cat lives in a scent-focused world and all creatures within his social group, and even inanimate objects, are anointed with his unique smell by means of scent glands in his face, body and tail. When your cat rubs round your legs to greet you, it is his version of a cat-to-cat greeting that would involve mutual rubbing of the face and body. As your face is a little too far away he will, for convenience, use your leg. Some cats really try to make the effort and stand on their hind legs to attempt a head butt on a body part as near to their owner's face as they can manage. Once you have been suitably rubbed, your cat will then take himself off to groom his body and check out your scent, so if you've been with another cat . . . he will know.

What does it mean when my cat sticks his bottom in my face?

On the 'greeting after an absence' theme, cats from the same social group will sniff one another as well as rub to get maximum information. A part of this includes a nose up the anus, so your cat is merely giving you the opportunity to check him out. Thank goodness we humans limit ourselves to shaking hands.

Why are non-cat people so attractive to cats?

This is not, as some suspect, a sign of the warped sense of humour of the average cat. It's all about body language. When a cat enters a room all the cat lovers start staring at the beautiful creature. They lurch towards him, extend their hands, rub their fingers and thumbs together in a 'look what I've got' gesture and make squeaky, kissy sounds with their mouths or do high-pitched baby talk. The non-cat person in the meantime is attempting to make her- or himself as inconspicuous as possible. She will probably be sitting extremely still, looking down and facing the opposite direction, keeping her hands firmly on her lap and maintaining absolute silence. If you were a cat, going about your normal business, confronted with this diverse display of human behaviour, which one would you gravitate towards as the least obnoxious?

Why does my cat put toys in his water; do cats kill by drowning?

Playing with toys is all about hunting, and eating is a natural extension of the same process. Natural consumption of prey is preceded by various stages from stalking to pouncing and tearing to chewing. This just isn't possible (or necessary) in the average domestic setting, but some cats still need that winning combination of sensations. So they chew at toys near their food and then eat a few mouthfuls of the gourmet paté that is not quite so chewy; they may even put toys in or near their food bowl under the same illusion. As many owners still put water and food together in one place for their cats I think that death by drowning is probably a red herring. It's the association with the food that triggers the action.

Why does my cat bring me a mouse?

This really isn't something anyone should take personally. There are all sorts of theories suggesting that it is a kind of gift, with the cat acting in a parental role and we are surrogate, if somewhat large, kittens. The reality is that sometimes things aren't all about us. A cat is, and always will be, a predator. We may selectively breed for the cute attributes they display indoors, but let any cat out and he will eventually do what comes naturally. Those cats that weren't taught by their mothers to kill efficiently may never get it quite right, but some manage, despite the lack of early maternal guidance, to become quite adept. Cats who fill up at home with food will potentially still hunt; appetite only really affects the level of enthusiasm in their endeavours. Bringing the prey back home to you indicates that your cat feels secure there and it's a safe place to leave stuff to eat now or keep for later. Occasionally you will look over the side of your bed at 3 a.m. and see your delightful little darling chucking a dazed mouse around and flinging it from paw to paw. This looks bad but such is the nature of the beast. We domesticate them selectively to tone down the whole 'catch and despatch' hunting prowess of the true carnivore and then complain when they don't kill their prey but prefer to torture it a little instead. If this has become a real problem for you (and the local rodent population), then consider the advice in Chapter 6 on hunting.

Worse still, why does my cat bring me half a mouse?
Sometimes your cat is a little hungrier than usual but can't quite manage a whole one!

My cat destroys the sofa yet his scratching post is pristine. Why?
I can't imagine what the person who invented the first scratching post was thinking. 'I know,' he said, 'let's design something for cats to scratch that's far too short, really wobbly and less than half as attractive as the average sofa!' Cats scratch to maintain their claws, exercise muscles and mark territory. Unfortunately our expensive armchairs and sofas are just at the right height for a cat to scratch at full stretch, heavy enough to resist even the most energetic pull and right next to where the average cat spends most of his time asleep. Once your cat has deposited his scent on your sofa, and left that telltale area of pulled threads, he will return to top it up on a daily basis until all the fabric has been destroyed and most of the stuffing has come out.

To stop this sort of damage (and also protect your carpets, textured wallpaper, divan beds and pine furniture) you need to provide acceptable rigid scratching posts of the right height and texture and put them in the sort of place where cats like to scratch – i.e., near where they sleep and in strategic places near windows, glass doors and entry and exit points. See Chapter 8 for more details.

Why does my cat roll over and show me his tummy yet scratch me when I tickle it?

This may seem like a case of mixed messages but it's we who get it wrong. When a cat greets his human companion he makes a display of trust by exposing his belly. This shows that he is prepared to put himself in a most vulnerable position when in the presence of his favourite human. Often this sort of signalling takes place if the cat is woken from a cosy sleep by our arrival. What your cat is really saying to you when he does this is that he feels safe with you in theory but his claws are a warning that he doesn't particularly want to put it to the test. So next time your cat rolls over and shows you his tummy, try to resist the temptation to fiddle; it will only end in tears.

My cat sits and stares at the cat flap. Can it really be entertaining?

Unfortunately this isn't your cat enjoying the feline version of television's *Play School*: 'Let's see what's through the square window this morning, shall we?' There is no way your cat will charge through the cat flap without thoroughly considering the wide variety of horrible fates that may await him. To us the cat flap is a convenient tool to enable the 'home-alone' cat to come and go as he pleases. To the cat it is the Portal of Hades and worse. Not only may the devil be outside but the wretched device may let the devil in as well! Your cat may be on sentinel duty, awaiting the arrival of something nasty, unable to tear himself away for fear that it will get in without his knowing.

Cats are territorial creatures and as such they are risk assessors extraordinaire. You may have observed your cat push his head through the flap and stay in that position while sniffing furiously. This enables him to check for ambushes and the aromas of cats that have passed by. What you won't see is your cat leave the cat flap with a spring in his step and gaily march across the middle of the garden without a care in the world. He will sit on the patio, using the potted plants for camouflage, or follow the contours of the building and nearby fences, which provide some protection. It's a dangerous place out there.

Why won't my cat play with me?
I would suggest that every cat will play if you press the right buttons. Play is an inherent part of the species make-up; it is hunting practice and great fun to boot, so there is absolutely no reason why your cat wouldn't play under the right circumstances. There are some questions that you need to ask yourself if you are frustrated that your cat won't play:

Am I giving my cat the right toys? A bunch of brightly coloured expensive nonsense from a pet shop stuck in a basket in the corner of the room won't necessarily comprise 'the right toys'. You should choose things that best mirror a cat's natural prey – for example, real fur toys of a similar size to the average shrew. Toys that move – e.g., anything dangling from a fishing rod that you can agitate in front of him – are hard to resist. Try lacing everything with a liberal sprinkle of catnip. Many cats find this incredibly exciting and it may prompt a game.

Is my cat poorly? Sometimes the first sign of illness that shows is a decline in what I would refer to as 'recreational activities'. Get him checked out by your vet.

Is my cat stressed? A cat really gives himself to the moment when he plays, so if he is on guard or feels threatened he will be reluctant to indulge in games for fear of being caught without a fixed bayonet, so to speak. If he stands in a corner and watches his cat 'pal' leaping around madly, maybe the two aren't such great 'pals' after all.

Some cats are naturally more playful than others, but there is no reason for your cat not to love a good game right up to his dotage. Admittedly the games are less boisterous then, but they are just as enjoyable.

My cat has a mad half-hour in the evening. What does this mean? This is usually the speciality of the house cat. It is perfectly normal and quite understandable when you recognize it for what it is. In the natural world outside cats would spend a great deal of time during the day stalking and chasing prey or fleeing danger in their adrenalin-fuelled lifestyle of hunting and hazard avoidance. A day in the life of the average house cat rarely includes anything dangerous and the energy doesn't get used up. Suddenly, often without warning, this energy will burst out and your cat will act out a little fantasy role-playing, alternating between the hunter and the hunted, dashing round the house with a flicking tail and widely dilated pupils. This often occurs at times of the day and night when cats are naturally more active – for example, at dusk – and it can be triggered by a loud noise, a visit to the litter tray or something quite inconsequential. Your cat may stare into the top corner of the room before launching himself across the carpet, but don't worry: he isn't having a psychic moment. He's just using up that excess energy. You can prevent these explosive escapades by stimulating your house cat as much as possible with games, toys and other activities.

My cat covers his food, and not his poo. Is this normal?
First things first. Some cats will scrape around their food and water bowls using the same action that is normally associated with toilet habits in the litter tray. There are a number of theories about this behaviour. One suggests that the cat, not wishing to consume the whole contents of his bowl, is attempting to cover it in order to return to it at a later time. This may be the behaviour of some big cats that catch large prey, but most of the food that the domestic cat eats comes in bite-sized chunks. Another theory has it that the cat is not finding the smell of the food particularly appealing and the covering is an attempt to cleanse the area and remove the offensive odour. If this scraping around becomes an issue for you, it may be worth giving your cat smaller, more frequent meals, or offering dry food, and then 'leftovers' will not represent a problem.

If, however, it seems ironic to you that your cat diligently attempts to cover his food yet leaves huge piles of faeces in the middle of your garden, then we still need to talk. Your cat may be 'middening'; that means depositing faeces as a marking gesture rather than a straightforward bowel evacuation. Is your cat up against some stiff competition in the garden from neighbours' cats? Does he feel he has something to prove regarding territorial rights?

It's important to make sure that there are safe, private areas in your garden that have loose soil or sand where your cat can toilet and cover with ease, as some acts of apparent middening are actually just cats objecting to rock-hard borders.

Why does my cat knead me; is it because he needs me?
When kittens are born they are totally dependent on their mother for their nutrition. When they suckle at their mothers' teats they tread with alternate front paws at her body to

stimulate the flow of milk. If your cat is kneading on your lap or chest when you have a cuddle, this shows that he associates the intense and secure nature of your relationship with that of his furry mother when he was a tiny kitten. Some cats will take the process one step further and dribble uncontrollably as they anticipate the milk feed that kneading usually produced.

Most cats grow out of this behaviour when it ceases to be necessary, but a few will retain it into adulthood, particularly when close by someone they really love. If your cat does this to you every now and then it's just a sign that you have a great relationship. If your cat cannot go a moment without seeking you out to dribble and tread, he may be a little needy. It's good when our cats love us, but we don't want them to be dependent on us. It wouldn't be a bad thing to deter him every now and then and encourage instead a rather more manly hunting game or a quick constitutional round the garden. Self-reliance is something to be encouraged in cats, for their own good.

Should I discourage my cat from sucking my ear lobes?

This is an extension of the kneading, suckling behaviour that has been taken to a slightly yucky level. The ear lobe is a perfect size to suck but I really can't understand why any owner would actively encourage this. It is OK to stop your cat doing it; I have to say that this is probably one of those things that is not acceptable in polite company. It does look a bit odd.

Should I worry if my cat chews cardboard?

Cats have evolved to kill and consume 'meat on the bone', so their teeth are highly developed and effective piercing and shearing tools. Many cats are given soft food or tiny biscuits to eat, and while this food is nutritionally excellent it does deny them the opportunity to experience the sensory buzz they would undoubtedly get when tearing open a mouse. If you surreptitiously watch your cat next time he 'chews' a box, you will notice that he does so with the large molars at the back of his mouth. You won't see him with bulging cheeks chewing the cardboard like a cow with grass; he is actually using the same sequence of behaviour that he would to cut through tough meat and bone before swallowing whole chunks. It may be redirected on to something without a pulse but it is, nonetheless, a highly rewarding experience for those cats that develop this habit.

Experts believe that the motivation for this behaviour is similar to that of wool-eaters, cable-chewers and cats with other pica preferences. Of all the substances to chew, cardboard is one of the safest, as it breaks down reasonably well in the gut should your cat take the behaviour one step further and swallow it too. It is probably better to sacrifice the odd cardboard box to indulge your cat's habit rather than risk the behaviour being directed on to something potentially more problematic. It tends to be the mark of the bored house cat and, often, the highly bred pedigree. An opportunity to hunt and do it for real may resolve the habit if that is a practical option.

Why won't my cat use his new cat bed?

Just because you spent a fortune on a new bed for your cat and it's made from leopard-print fake fur does not necessarily mean that he will like it. Cats need to have a sense of control over their environment and that includes when and where they choose to sleep. If you investigate your cat's current favoured places to rest you may get a clue why he rejects his posh new beanbag. Many of the locations he chooses for sleep are raised – for example, the sofa or your duvet – so putting the new bed on the floor with a gale-force draught whistling past it may not be quite so appealing. I succumbed to the lure of a designer bed for Mangus, only to find she politely rejected it in favour of the sofa, on the principle that 'if it's good enough for you, it's good enough for me'. I personally think fancy fleece blankets are the answer; you can position them on favourite chairs or beds and your cat will probably be more likely to curl up directly on them and use them continuously.

If you are now stuck with a brand-new cat bed that has never been christened, try placing it on a raised surface in the room that your cat favours and near a source of heat – for example, a radiator. Sprinkle it with a little catnip (if your cat is appreciative of its charms) and then ignore it completely. With any luck your cat may deign to sit in it from time to time.

Why doesn't my cat like our lovely garden?

Your idea of a lovely garden may not be your cat's. There is a trend towards decking, patios and feng shui at the moment and many house owners with small back courtyards or gardens convert their space to an open area of peace and tranquillity. Not so for their cats, I'm afraid. Cats need camouflage, they need high perches in the garden to observe their territory and they need grass. The ideal garden for your cat is probably the one you had when you first moved into your home – what the estate agent described as 'overgrown and in need of landscaping'.

If you are keen to give your cat what he really wants in order to encourage him to spend time in the fresh air, then put around the edges of the space plenty of pots with bushy plants or shrubs to give him some hideaways. If you would like to devote some space to your cat, look at Chapter 6 for more ideas on the cat-friendly garden.

My cat grimaces after he sniffs something. Does this mean he doesn't like it?

This is an extraordinary expression that cats assume when they are particularly interested in a smell. They approach the odorous object in question and sniff, but then open their mouths and sit there for a while, staring into the middle distance. This is called the flehmen reaction; it is not attractive but a necessary thing in the scent-biased world that our felines inhabit. Cats have an extra organ of scent, called the vomeronasal or Jacobson's organ, that has an opening behind the incisors on the roof of the mouth. This additional piece of equipment enables cats to 'taste' smells and get every little bit of information out of them. It is particularly used for urine, as this contains a complex concoction of messages about the cat that passed it.

What does it mean when my cat purrs at the vet's surgery?

There has been a degree of mystery associated with the cat's purr; all smaller felids – including the serval, puma and ocelot – and even some large cats, such as lions and cheetahs, purr. As purring takes place when kittens suckle or when cats are stroked, there was an assumption that it signified contentment. However, we now know that cats also purr when they are frightened, in pain or even dying; this would probably account for the cat purring furiously while being handled by the vet.

There are a number of theories why cats purr, including signalling receptivity to social contact and self-reassurance. The theory I am particularly interested in suggests that the purr has healing properties. Cats purr at a frequency that is considered therapeutic for bone growth, fracture healing, muscle growth, pain relief, wound healing and mobility of joints. This would provide a wonderful explanation for the purr of a seriously injured cat; it's a theory definitely worthy of further investigation.

Is my cat happy when he wags his tail?

If you see your cat wagging his tail do not be fooled into believing it has the same positive connotations as it would if he were a dog. A highly excited cat may thrash his tail from side to side just before he pounces on something thrilling. A cat in pain may wag his tail slightly and a cat about to show aggression may also do so. It doesn't, however, indicate anger directly; the aggression is often a consequence of the emotional conflict that the cat is experiencing. A wagging tail is a sign of a cat caught between two actions – for example: 'Should I go in for dinner or chase that mouse again?'

Why does my cat sit in high places?

A cat's behaviour is greatly influenced by its territorial nature and survival instincts. Cats achieve a degree of security when sitting on a high platform, or up a tree or on a shed, as they are in a good position to view the territory without being in immediate danger. Cats kept exclusively indoors often instinctively seek out a high place to escape danger, so it's really important to provide appropriate opportunities for them to do so – for example, a wardrobe or the top of the kitchen cupboard. It's probably not wise to attempt to coax your cat down, as he is there for a very definite reason and you risk distressing him.

Why doesn't my kitten miaow?
Vocalization is a very individual skill and cats will develop, over the years, a range of sounds specific to the relationship with their owners. Some cats are naturally more talkative than others – for example, the Oriental breeds and my own cat Mangus rarely stop shouting. If you want to encourage more vocalization from a previously mute kitten or cat, it's important to reward every tiny noise that emanates from him with a treat that he finds particularly appealing.

Why does my cat wake me up at night?
Cats have evolved to hunt at times when ambient light is low, so they are naturally more active at night. Some domestic cats adapt well to our more diurnal pattern of wakefulness and get used to sleeping when we do. Most, however, still have quite a bit of get-up-and-go when the house is at its quietest. If you are a very responsive and attentive owner, why wouldn't your cat wake you for a quick game or even a snack to while away the hours? Some owners tie themselves in knots trying to accommodate the nocturnal demands of their pampered cats. My best advice, to prevent ill health as a consequence of your sleep deprivation, is to give your cat attention during the day and ignore him completely at night. If, though, the behaviour coincides with other significant behavioural changes, such as intense vocalization, or your cat is elderly, then I would suggest a check-up at the vet's, as it could be an early warning sign that your cat is unwell (see below).

Why does my cat call at night?
If your cat isn't a female in season, there are several reasons why he might do this. He may be bored and full of energy; some cats will rush around and vocalize at night as a form of

frustrated behaviour. Calling at night can also be one of the first signs of hyperthyroidism or even senility in older cats (see Chapter 9). Vocalizing is often a precursor to urine-spraying, so your cat may be responding to a potential feline intruder outside in the garden. If your cat is calling at night it is wise to attempt to establish his reason for doing so before following the standard recommendation, 'If all is well ignore it'!

What does it mean when my cat's tail goes bushy?

A cat has the capacity to fluff up his tail and the fur along his back to stand erect at a right angle to the skin, referred to as pilo-erection. This gives the cat a much larger silhouette and, together with an arched back and a sideways stance, is used to signal defensive aggression. Some cats that experience a sudden fright will instinctively puff up their tail before investigating the perceived danger a little further.

Why does my cat drink from the tap?

Cats' motivation to drink is not connected to hunger, so many find it confusing when water bowls are provided directly adjacent to their usual feeding area. Some cats adapt to this strange set-up relatively easily, but others reject this water as unsuitable and seek other more acceptable sources. Taps, glasses of water, vases and goldfish bowls are all potential thirst-quenchers, but your best option is to provide dedicated drinking vessels in alternative locations well away from your cat's food. Some cats also prefer running water, so a pet drinking fountain can be purchased as a more practical source than a constantly dripping tap.

Why does my cat love my husband's armpits?

Some men find this a hilarious party piece and actively

encourage their cats to perform this little ritual. Armpits have scent glands that produce pheromones and sexual aromas. Some cats are extremely attracted to this smell and they will sniff, lick, nuzzle and rub in complete ecstasy while, often, drooling profusely. Unfortunately the smells of clothing that have been worn adjacent to the armpits and genitals are virtually irresistible for some, so thrusting the cat away from the actual body part isn't always a cure for the problem. If you are a person who isn't terribly keen on your cat doing this, then ensure all dirty clothing is put in the laundry basket, not discarded on the floor, and, whenever your cat approaches that part of your body with a glint in his eye, gently get up and walk away. Improving your personal hygiene won't make the problem go away; that smell is still there for your cat no matter how clean you are.

Why does my cat make love to my dressing gown?

Another less attractive habit of the neutered male cat is 'humping' items of worn clothing or unsuspecting teddies. The behaviour often includes gathering the garment together into a heap or grabbing the toy by the mouth; the front paws then tread up and down and the cat purrs. He then pushes the object between his back legs and starts to thrust his pelvis backwards and forwards rapidly with a far-off gaze in his eyes. It is best not to interrupt at this stage as he will continue to the bitter end and then remove himself from the object of his desire and lick his penis distractedly for several moments afterwards.

This behaviour, masturbation by any other name, is triggered by a texture or scent and tends to recur only when exposed to one particular object. Some cats generalize to other similar objects or even, if you are very unlucky, to human arms and legs. This is probably when you should seek professional help from a cat-behaviour counsellor to address this, as it's a habit that is hard to live with. It is considered to be a displacement activity or coping strategy for something that is stressing your cat, so it may be a symptom of an underlying problem.

APPENDIX 2

My Cat's Got What?!

UNDERSTANDING VETERINARY TERMINOLOGY AND DIAGNOSES

VISITS TO THE VET CAN BE EXTREMELY STRESSFUL AND NOT just for your cat. If our furry friends are ill, there are so many emotions rushing through us that it's hard to take in the details of the tests and the treatment, let alone the name of the actual disease. I have had countless calls from past clients who have asked for a quick interpretation of their cat's ailments. I'm not really in the best position to do this but I try to give people strategies to get the information they need – for example:

- Take someone with you to the vet if you are expecting bad news; two heads are better than one at absorbing all the information.

- Ask your vet to write down all the relevant things you need to know.
- If you are still unsure, make another appointment to see your vet, without your cat, and take a notepad so that you can jot things down.
- Don't be afraid to ask questions; your vet is there to help you!
- If your vet is finding it difficult to make a diagnosis of your cat's condition, don't be afraid to request a second opinion or referral to a specialist. It's standard procedure under these circumstances.

It may be useful for you to have a list of the fifty most common diseases, injuries and symptoms, along with a glossary of terms to guide you through the minefield of veterinary terminology. Please don't use this as your sole source of information regarding a particular complaint if you want to know as much as possible; it's merely intended to be an initial stress-buster if you feel bogged down by all the vet-speak. There are a number of excellent websites (see the list at the back of the book) that will give you good, accurate and up-to-date information about various conditions and their treatments. If you intend to trawl the Internet, please be warned: many sites look professional but give bad advice. You don't want to add stress to your situation rather than relieve it.

Glossary of veterinary terminology

Abdomen – the region of the body between the chest and the tail

Allergen – any substance that can induce allergies

Analgesia – a temporary absence of the sensation of pain; induced by drugs
Antihistamine – a drug that reduces acute inflammation, especially in allergic reactions
Anorexia – complete loss of appetite
Antipyretic – a drug that reduces fever
Aorta – the largest artery leading from the heart to all parts of the body
Ascites – a build-up of fluid in the abdominal cavity
Aural haematoma – an accumulation of blood in the tissues of the ear that clots to form a hard swelling
Axillae – the area of the body equivalent to the armpits
Azotaemia – an increase in the waste nitrogen levels in the circulation, measured as blood urea (BUN) and creatinine, indicating kidney damage/failure
Benign – used to describe a tumour that does not invade and destroy tissue or spread to other parts of the body
Biopsy – the removal of usually small pieces of body tissue to identify specific disease processes
Bronchitis – inflammation of the airways in the lungs
Bronchodilator – a drug or agent that dilates the airways
Cannula – a hollow tube inserted into a bodily cavity, using a fine metal rod to guide it, to drain fluid or insert medication
Cardiac – referring to the heart
Caudal – in the direction of the tail
Congenital – present at birth
Conjunctiva – the mucous membrane that lines the inside of the eyelids
Contagious – infectious, referring to a disease spread by direct animal-to-animal contact
Contrast radiography – X-rays using a substance (called a contrast medium) that increases the visibility of an internal structure

Cornea – the transparent part of the front of the eyeball
Cyanosis – bluish discoloration of the skin and mucous membranes caused by lack of oxygen in the blood
Diaphragm – a dome-shaped divider made up of muscle and tendons that separates the abdomen from the thorax
Diuretic – a drug or other substance that increases urine production and hence water loss from the body
Dyspnoea – laboured or difficult breathing
Dysuria – difficulty in passing urine
ECG (electrocardiograph) – the recording of the heart's electrical activity to diagnose disease
Enteral – relating to the intestinal tract
Epiphora – overflowing tears
Fluid therapy – the administering of fluids into the body to replace those lost due to disease or other circumstances
Gastrointestinal – relating to the stomach and intestines
Gingivitis – inflammation of the gums
Haematuria – blood in the urine
Heart murmur – a noise heard via a stethoscope that is generated by the turbulent flow of blood through the heart or blood vessels caused by cardiac damage or defects
Hepatic – relating to the liver
Hypertension – high blood pressure
Hypoglycaemia – abnormally low levels of glucose in the blood
Hypothermia – abnormally low body temperature
Infectious – referring to disease caused by harmful substances – e.g., bacteria invading the body
Intravenous – into the vein
Jaundice – a yellow discoloration of the body tissues caused by an accumulation of bile pigments associated with liver disease
Keratectomy – surgical removal of part of the cornea

Lymphatic system – a circulatory system in the body that filters out bacteria and other foreign substances
Lymph nodes – parts of the lymphatic system
Metastasis – the process by which tumours spread to other parts of the body not connected with the site of the original mass
Mucous membrane – a moist layer of tissue that lines hollow organs – e.g., the respiratory tract
Mycoplasma haemofelis – an infectious disease causing fever, anaemia and rapid weight loss
Nasopharynx – the part of the pharynx lying behind the nasal cavity
Non-steroidal anti-inflammatories (NSAIDs) – a large group of drugs that reduce temperature, inflammation and pain
Ocular – relating to the eye
Oedema – the accumulation of excess fluid in the tissues or a body cavity
Packed cell volume (PCV or haematocrit) – the proportion of the blood volume occupied by cells, used to diagnose dehydration or anaemia
Pathogen – a substance capable of causing disease
Perineal – relating to the region surrounding the anus and genitals
Pharyngitis – inflammation of the pharynx (back of the throat)
Pleural effusion – a fluid build-up between the chest wall and the lungs (the pleural cavity)
Pollakiuria – urinating more frequently than normal
Polydipsia – excessive drinking
Polyuria – the production of abnormally large quantities of urine
Pruritis – itchiness
Pyrexia – high temperature

Renal – relating to the kidney
Retina – the light-sensitive layer that lines the back of the eye
Sclera – the white part of the eye
Serum biochemistry – analysis of blood serum to detect the presence of disease
Soft palate – the soft part at the back of the roof of the mouth
Subcutaneous – beneath the skin
Thoracocentesis – the aspiration of material – for example, air or fluid – from the pleural cavity using a hollow needle inserted through the chest wall
Thorax – the chest
Ultrasound – a technique using ultrasonic (sound) waves that enables internal structures of the body to be viewed
Urinalysis – examination of urine to detect disease
Uveitis – inflammation of the vascular, pigmented layer of the eye

A layperson's guide to veterinary diagnoses

These are probably the most common diseases, symptoms and injuries that your cat may experience. I have included basic treatment protocols and prognoses, but please ask your vet for a more personal overview of your cat's situation. The definitions for all the medical terminology can be found in the Glossary.

1 *Allergic flea dermatitis* – also known as miliary eczema. Symptoms include pruritis, scaly skin, fur loss, over-grooming and secondary bacterial infection. It is caused by an allergy to flea saliva when the flea bites. Some cats are so sensitive that a single bite can set off a massive itchy reaction. Treatment includes regular topical flea-control

products from your vet; he or she may advise using them more frequently than the manufacturer recommends. (See Chapter 3 on fleas!) You will also need to treat your home thoroughly against fleas, paying particular attention to dark places under furniture and curtains. The vet may also give your cat steroids, to reduce the itch, and possibly antibiotics if the skin lesions have become infected. Antihistamines too are used for long-term allergies. The prognosis is good with excellent persistent flea control – irrespective of the season.

2 *Anaemia* – reduced circulating red blood cells diagnosed by packed cell volume (PCV or haematocrit). Cats suffering from anaemia also have pale mucous membranes and may be lethargic and even panting if the problem has an acute onset. Causes are many and varied, including blood loss and destruction of blood cells owing to disease. Diseases that may cause anaemia include mycoplasma haemofelis, kidney disease, FIV, FeLV, tumours and poisoning. Vets will explore whether or not your cat is producing more red blood cells to replace those that have been lost (regenerative anaemia) or not (non-regenerative anaemia). Treatment depends on the underlying cause, but symptomatic treatment may include fluid therapy or blood transfusions. The prognosis for anaemia depends on the cause; unfortunately symptomatic treatment will not be effective on its own in the long term.

3 *Arthritis* – cats suffer specifically from osteoarthritis, or inflammation of the joints, which causes pain and often results in restricted mobility. The joints affected are usually the hip, elbow and spine. It is still an under-diagnosed

problem but recent research suggests that over 65 per cent of cats over the age of twelve suffer from this condition. Even on X-rays it is difficult to diagnose many cases. Arthritis is often identified by observing the cat's symptoms and response to treatment (see Chapter 9 for details). Treatment includes NSAIDs, 'nutraceuticals' (for example, glucosamine and chondroitin) and manipulation (for example, chiropractic and acupuncture). Weight loss in the obese sufferers can have a very positive effect on their overall pain levels. Prognosis depends on the severity of the problem and patients usually require ongoing treatment for life.

4 *Asthma* – also known as allergic bronchitis. This is more common in Siamese but can affect all cats. Symptoms include bouts of coughing, panting, cyanosis and even collapse. Ongoing treatment for the condition includes bronchodilators, steroids (administered via an inhalation mask specially designed for the cat), oral steroids and reduction of potential allergens – e.g., house dust or cat litter dust. Acute episodes require emergency treatment, including oxygen therapy and intravenous medication. The prognosis is fair although the cat will inevitably require long-term treatment.

5 *Blocked anal glands* – the anal glands (or sacs) are situated at 'twenty minutes to four' on the cat's bottom! They contain fluid that normally discharges on to the cat's faeces when it defecates. Occasionally this fluid can become thickened and infected or the cat's motions become loose and they will not evacuate. The fluid then builds up and causes discomfort. Symptoms include 'scooting', excessive

grooming around the anus and even toileting problems. Treatment includes expressing manually, or flushing using a fine cannula inserted into the gland while the cat is sedated. Antibiotics and analgesics may also be prescribed. The prognosis is good but your cat may need regular expression of the glands or dietary changes to improve the consistency of the stools.

6 *Blocked tear duct* – the cat has tear ducts in the rim of the eye that drain into the nasal cavity. In some cats, particularly pedigrees with non-standard faces, these ducts are anatomically unusually narrow or they become blocked due to infection. Diagnosis is by using fluorescent dye in the eye that, if the tear duct is blocked, will run down the face or, if functional, will run out of the nose. If the blockage affects one eye only and is of recent onset, this could indicate a tumour or foreign body. Treatment isn't appropriate if the problem is anatomical, but various cleansers are available to wipe over the fur surrounding the eye that tends to become heavily stained as a result of the overflowing tears. Ongoing maintenance is necessary if the problem is anatomical.

7 *Bordetella* – *bordetella bronchisepta*, an organism that is naturally occurring in the cat's upper respiratory tract, occasionally overgrows and causes symptoms such as coughing, pyrexia and pharyngitis resulting in anorexia. It can be vaccinated against; this is appropriate in households with large numbers of cats, as it appears to be contagious. Treatment depends on the severity of the symptoms and includes antibiotics, analgesia, antipyretics (NSAIDs) and fluid therapy. The prognosis is good as cats usually recover

unless the individual is severely immuno-compromised or there are concurrent problems.

8 *Cat-bite abscess* – cats have some very nasty bacteria on their teeth and when one bites another the canines puncture the skin and enter the underlying tissue, depositing this horrible concoction. Once underneath it cannot escape and is recognized by the body's protective system as an enemy. White blood cells attack the invading bacteria and the body encapsulates it. Cat-bite abscesses (CBAs) vary in severity, depending how deeply they penetrate and where they occur. Cats suffering from CBA will be lethargic, pyrexic, anorexic and, depending on the bite's location, lame or in severe pain. There may be noticeable and rapid swelling in the affected area. Some CBAs burst, releasing malodorous pus and affording tremendous relief to the cat. Others need veterinary intervention that may involve lancing and draining of the abscess fluid together with analgesia and antibiotics. The prognosis is excellent but can be a problem if the CBA is located on a limb joint or penetrating the abdomen or thorax. CBAs are also a likely means of transmission for FIV and FeLV (see below).

9 *Cat flu/herpes/calicivirus* – cat flu is caused by a number of viruses, including herpes, calicivirus and even bordetella. Symptoms include anorexia, owing to nasal discharge and blockage, ocular discharge and ulceration of the tongue. It can be vaccinated against but this will not protect cats from all strains of the virus, although symptoms are usually milder. Treatment is symptomatic and may involve fluid therapy, enteral feeding, antibiotics (against any secondary infection), analgesia and antipyretics. The prognosis varies

enormously as recovery and treatment may be prolonged while the cat remains contagious. Recovered kittens may remain carriers and have chronic upper respiratory signs – i.e., 'snuffles' with a thick nasal discharge.

10 *Chlamydophila felis* – chlamydophila is an organism that causes conjunctivitis (see below), often affecting one eye. It can also cause respiratory symptoms and abortion in pregnant queens. It is contagious but species-specific and cannot transmit to humans, so don't panic. Diagnosis is made via a swab of the eye or nasopharynx. It can be vaccinated against; this is particularly popular in breeding establishments, for obvious reasons. The prognosis is variable as it can be difficult to eradicate from a 'colony' and even from the individual. Treatment includes a particular group of antibiotics called tetracyclines in the form of eye drops.

11 *Cholangiohepatitis* – this disease is due to inflammation of the liver and associated bile ducts. Symptoms include anorexia, depression, lethargy, jaundice, pyrexia and vomiting. Diagnosis is made by serum biochemistry, showing the presence of raised liver enzymes – e.g., ALT and ALKP – although definitive diagnosis should include biopsy of the liver. Treatment may include antibiotics, steroids and other supportive therapy. Prognosis is very varied, ranging from grave to very good, as the disease can wax and wane.

12 *Conjunctivitis* – inflammation of the conjunctiva causing pain, squinting, rubbing, discharge and reddening. It can be a symptom of disease or a consequence of a foreign body in the eye. Treatment includes removal of any foreign body,

antibiotics and analgesics. The prognosis is generally very good.

13 *Corneal ulcer* – if the cornea of the eye is damaged – for example, by a cat scratch or other trauma – the surface will ulcerate. Ulceration can also occur as a result of the herpes virus. Symptoms include squinting, pain, epiphora and reddening of the sclera. Diagnosis is via a fluorescent stain dropped on the cornea that will highlight areas of ulceration. Treatment includes antibiotics and close monitoring. For severe or unresponsive cases the vet may perform a keratectomy under general anaesthetic or protect the eye by suturing the third eyelid across or fit a contact lens. The prognosis is good assuming the eye heals normally.

14 *Detached retina* – if the retina completely detaches from the back of the eye it will lead to sudden blindness. Detached retina is a potential consequence of various diseases causing hypertension, including chronic renal failure and hyperthyroidism. Symptoms include permanently dilated pupils and behavioural changes – for example, staying indoors more and increased vocalization. If you are able to identify this condition rapidly (within a couple of days) and the hypertension is addressed, the retina may re-attach but the sight will probably still be lost; therefore the prognosis, for that alone, is poor.

15 *Diabetes* – glucose levels in the body are controlled by a hormone, produced in the pancreas, called insulin. In cats with diabetes mellitus either insulin stops being produced or the body no longer recognizes the insulin. This results in

an increase in glucose levels within the body, causing polydipsia, polyuria, increased appetite, weight loss and occasionally cystitis. If the condition remains untreated it can lead to anorexia, kidney failure and collapse. Diagnosis involves blood tests and urinalysis. Treatment includes regular insulin injections and special diets. Some cats will respond to treatment and some will spontaneously recover. Others, unfortunately, are very hard to stabilize and often have concurrent illness.

16 *Diaphragmatic hernia* – this occurs in extreme trauma cases, usually road-traffic accidents, in which the diaphragm ruptures and the abdominal contents push through to the space around the lungs. Symptoms include dyspnoea, abdominal breathing (body rising and falling rather than the chest), anorexia and lethargy, though this condition rarely occurs in isolation. Diagnosis is through X-ray, but these patients are often too sick to have radiographs taken. Treatment includes initial stabilization, including fluid therapy, analgesia and oxygen. Ultimately the treatment is surgery, but this involves complicated anaesthesia and careful monitoring. The prognosis depends on the severity of the case and any concurrent trauma. Unfortunately, despite everyone's best efforts, patients often don't recover from the anaesthetic of the remedial surgery.

17 *Ear mites* – these are microscopic parasites that live in the ear canal and cause irritation and excessive wax production. Symptoms include head shaking, dark brown wax, scratching and associated infections. The head shaking can be so severe that it causes aural haematoma. Diagnosis is usually based on clinical symptoms, although small

particles of wax can be examined under a microscope to show the presence of *Otodectes cynotis*, the parasite in question. Treatment is via ear drops, administered according to your vet's instructions. The prognosis is excellent but the condition is contagious, so this may be a complication in a multi-cat household.

18 *Ear polyps* – these are usually benign growths that occur in the ear canal. Symptoms include ear irritation, head shaking and infection (usually causing a foul smell and discharge). Treatment includes antibiotics, either orally or in the form of drops in the ear, and anti-inflammatories. If the problem is recurrent over a period of time, the treatment of choice is a surgical technique involving partial or complete removal of the ear canal. The prognosis is variable depending on the severity of the condition.

19 *Entropion* – this can be a congenital or acquired condition that causes the eyelids to fold inwards, irritating the cornea and causing ulceration. Symptoms include epiphora, squinting, pain and reddening of the sclera. Treatment should address the underlying cause, and surgical reconstruction of the eyelid may be necessary. The prognosis is usually excellent but may involve several surgical procedures, as the surgeon will err on the side of caution when it comes to the amount of tissue removed on each occasion to create the necessary effect.

20 *Eosinophilic granuloma complex* – EGC is an immune-mediated condition that results in the formation of raised plaques on the lip, tongue or palate. Symptoms include excessive salivation and difficulty in eating. The condition

is usually associated with pruritis and over-grooming. Diagnosis is by visual examination and biopsies. EGC can be difficult to treat, but any concurrent skin allergy should be addressed to get the best results. Treatment includes flea control, a novel protein diet and steroids. The prognosis depends on the elimination of any contributory conditions.

21 *Feline idiopathic cystitis* – this is a complicated condition of the urinary tract that is considered to be stress-related and results in inflammation of the bladder wall. Symptoms include pain, dysuria, haematuria and even, in the male cat, a complete inability to urinate. Other more vague symptoms include inappropriate urination and over-grooming of the lower abdomen. Diagnosis is usually by the exclusion of other possible diseases, urinalysis and scrutiny of the cat's presenting symptoms. Treatment in the short term is ineffective, as the condition is self-limiting; the causes are complicated and multi-factorial, but many vets give antibiotics in the first instance. The best treatment for this condition is now considered to be increased fluid intake to dilute the urine and behaviour therapy to reduce the cat's underlying stresses. The prognosis is highly dependent on the severity of the symptoms and the ability to address the underlying stress trigger.

22 *FeLV* – feline leukaemia virus is a viral condition that affects the immune system and causes a wide variety of symptoms and associated conditions. It can be vaccinated against. The virus transmits from cat to cat via blood and saliva – e.g., cat bites – and from queen to kittens. It tends to cause tumours within the lymphatic system or bone marrow and anaemia. Diagnosis is via a dedicated blood test. There is

no cure for this condition and each patient is treated symptomatically. The prognosis is grave and euthanasia may be recommended.

23 *FIP* – feline infectious peritonitis is a fatal disease caused by a type of virus called coronavirus. It is an extremely complicated disease that is still poorly understood. Symptoms are wide and varied, including ascites, pleural effusion (see below), pyrexia, anorexia, diarrhoea, jaundice, uveitis and even seizures. A definitive diagnosis can only be made by tissue analysis after the cat has died, but presumptive diagnoses may be made in cases that present a raised coronavirus level *and* relevant symptoms. It's really important to note that an elevated coronavirus level in itself is not diagnostic of the condition. Treatment is palliative as, sadly, there is no cure for this condition.

24 *FIV* – feline immunodeficiency virus is caused by a similar retrovirus to HIV, although it is not transmissible to man. Cats can infect other cats through biting and from queen to kittens. It causes immuno-suppression in cats; symptoms include pyrexia, anorexia, gingivitis, diarrhoea, skin diseases and many other chronic illnesses. Diagnosis is via a dedicated blood test. The treatment is palliative or your vet may recommend immuno-therapy. There is no cure for this condition; however, cats may live a long, happy life after contracting FIV. Owners of FIV-positive cats are usually advised to keep them indoors to prevent further spread of the disease.

25 *Feline odontoclastic resorptive lesion* (FORL) – this condition, which used to be referred to as a neck lesion, involves a

hole developing in the bottom of the tooth, at the gum margin, that exposes the nerve. Symptoms include pain, anorexia, gingivitis and drooling, although many sufferers show no symptoms. Diagnosis is made via dental radiographs and visualization of the lesions during dental examination. Treatment is extraction of the tooth under general anaesthetic; some complicated cases may need a referral to a dental specialist. The prognosis is excellent if treated appropriately.

26 *Food hypersensitivity* – this can manifest itself as an itchy skin condition, over-grooming, vomiting or diarrhoea. Diagnosis is by exclusion of other possible differential diagnoses and, if it is suspected that it may be food hypersensitivity, the vet may recommend specialized blood tests or exclusion diets that contain novel sources of protein and carbohydrate. The specific allergen can be identified if symptoms return when that item is reintroduced into the diet. This process can take a long time and sometimes management can only be achieved through steroid administration. The prognosis depends on successful identification and avoidance of the allergen.

27 *Foreign bodies* (nasal – reverse sneezing) – cats rarely get gastrointestinal foreign bodies, although those cats with a wool-eating habit (or similar) are more at risk. As cats eat vegetation as part of their natural diet and then vomit, a part of the grass matter can get lodged between the soft palate and the pharynx or in the nose itself. Symptoms include 'reverse sneezing' (sounds like snorting), nasal discharge from one or both nostrils, irritation, swallowing/ gulping and general signs of distress. Diagnosis and treat-

ment are made by examination under general anaesthetic and removal of the offending object. The prognosis is excellent once it has been removed.

28 *Fractured mandibular symphisis* – this is a common occurrence in road-traffic accidents. The lower jaw splits in the middle, causing dramatic misalignment and associated problems with eating. Treatment involves initial stabilization of the patient and then surgical repair of the fracture, using wire, under general anaesthetic. Feeding tubes may be put in place at the same time to support the patient during the period of recovery. The prognosis is excellent providing there are no other concurrent problems.

29 *Giardiasis* – *Giardia sp* is an organism that lives in the small intestine and causes diarrhoea and weight loss, especially in kittens and young cats. Diagnosis is by specialist faecal analysis. The treatment includes antibiotics or worming solution as a prolonged course.

30 *Gingivostomatitis* – inflammation of the gums and mucous membrane lining the mouth. This may be related to tartar on the teeth but can also be a primary condition of unknown origin. Symptoms include severe pain, anorexia, drooling, halitosis and bleeding gums. Diagnosis is based on clinical examination and biopsies. Treatment includes steroids, analgesics, antibiotics, Interferon (an immune modulator) and, in severe cases, even total teeth extraction. The prognosis varies; some cats require life-long treatment to manage the condition.

31 *Hepatic lipidosis* – this occurs when fat infiltrates the liver. It

is usually seen in overweight cats that suddenly stop eating, probably as a result of underlying illness. Symptoms include jaundice, anorexia, lethargy and depression. Diagnosis is made through blood tests and ultimately liver biopsy. Treatment includes nutritional support, by feeding tube if the cat is unwilling or unable to eat, appetite stimulants and fluid therapy. The prognosis depends on the underlying cause for the initial loss of appetite.

32 *Hyperthyroidism* – this is a condition that affects predominantly older cats. A tumour, usually benign, develops on the thyroid glands situated either side of the throat, which causes an increase in the cat's metabolism. Tumours can be present individually or on both glands. Symptoms include increased appetite, weight loss, poor coat condition, diarrhoea and rapid heart rate. One of the first behavioural signs is often harsh vocalization at night. If the condition is left untreated it can lead to heart failure. Diagnosis is via a blood test; the vet may also be able to palpate the tumour in the cat's neck. Treatment includes drug therapy to stabilize the thyroid function, surgery to remove the tumour, or radioactive iodine therapy, although this last treatment option requires long periods of isolation for the patient. The prognosis is very good, unless there are other complications, although drug therapy has to be administered for life and tumours can reoccur in surgical cases.

33 *Hypertrophic cardiomyopathy* (HCM) – this is a condition that causes thickening of the wall of the heart. Symptoms include fluid developing in or around the lungs, causing tiredness, breathlessness, loss of appetite and general

malaise. Sometimes the first sign of a problem is development of a thromboembolism (see below) or cats may die suddenly with no apparent warning. Affected cats usually have an increased heart rate and sometimes have a heart murmur or abnormal heart rhythm. Diagnosis is by ECG, ultrasound and X-rays. This condition can be managed with diuretics and heart medication but fluid may need to be drained from the chest. The prognosis varies but can be poor.

34 *Inflammatory bowel disease* – IBD is a complicated condition caused by various triggers, including pathogens, stress or hypersensitivities. Symptoms include diarrhoea, weight loss, vomiting and general loss of condition. Diagnosis is by the elimination of other possible causes and ultimately bowel biopsies. Treatment includes identification and removal of triggers, steroids and metronidazole (a type of antibiotic that works as a bowel anti-inflammatory). The prognosis is good as the condition can be managed with a single course of treatment, although long-term medication may be necessary in some cases.

35 *Lymphoma* – this is cancer of the lymph nodes. It can affect any of the lymph nodes in the body but is readily seen when it is affecting the submandibular lymph nodes (under the chin), the popliteal lymph nodes (in the hind legs) and the prescapular lymph nodes (between the forelegs). There are also lymph nodes in the abdomen and thorax. The lymph nodes become enlarged and other symptoms can include anorexia, depression and weight loss. Diagnosis is by taking biopsies of one or more lymph nodes. Treatment may involve administration of steroids or chemotherapy but is not curative, although remission for a period of time

may be possible. In the long term a decision to euthanase will invariably have to be made.

36 *Malassezia* – this is a type of yeast that occurs naturally on the skin that, on certain individuals, will overgrow. Symptoms include pruritis, reddening of the skin, greasy areas and hair loss through over-grooming. The affected sites are usually the feet, axillae and groin. Diagnosis is made via blood tests or examination of the secretions under a microscope. Treatment includes regular bathing with medicated shampoo and oral medication. Malassezia can also be a secondary condition as a result of pruritis and resulting over-grooming that enables the yeast to overgrow. The prognosis is good if managed appropriately.

37 *Megacolon* – this is a condition that causes constipation due to inactivity of the colon. Symptoms include unproductive and painful straining, swelling of the abdomen, anorexia and inappropriate defecation outside the litter tray. Diagnosis is based on symptoms and elimination of other causes. Treatment includes dietary (increased fibre), enemas and oral medication. The condition cannot be cured but is usually manageable.

38 *Pancreatitis* – this is inflammation of the pancreas. Symptoms include anorexia, hypothermia, depression and weight loss. Diagnosis is by expensive specialized serum biochemistry, ultrasound and biopsy. Treatment includes supportive therapy – e.g., fluids and anti-sickness drugs. The prognosis varies according to the age and general health of the patient and many other factors.

39 *Pelvic fractures* – these commonly occur as a result of traffic accidents and falls. Symptoms can include inability to walk, tail paralysis, haematuria, pain and inability to urinate or defecate. Pelvic fractures are often associated with other injuries. Diagnosis requires X-rays. Treatment will depend on the location of the fracture within the pelvis. If there is significant narrowing of the pelvic canal, this may cause problems in defecation requiring medication to soften the stools. If the weight-bearing parts of the pelvis are affected, then surgery may be required. Many pelvic fractures will heal with six to eight weeks of cage rest. However, some pelvic fractures are so severe that even with surgery the prognosis is poor and euthanasia is necessary. Likewise if there is damage to the bladder nerves, and the cat is unable to urinate, treatment is not feasible.

40 *Periodontal disease* – the build-up of plaque and tartar on the teeth results in erosion and inflammation of the gums and loosening of the teeth. Symptoms include halitosis, reduced appetite, unkempt coat and discomfort on eating. The vet will diagnose this condition on oral examination. Treatment includes general anaesthetic, descaling and extraction of severely affected teeth. A course of antibiotics is also usually given. If multiple extractions are performed, analgesics will also be prescribed. This condition will re-occur in time if the cat remains on the same diet and no preventative dental care (e.g., tooth brushing, dental kibble) takes place, although some cats seem to be more prone than others.

41 *Pleural effusion* – this is an accumulation of fluid surrounding the lungs that causes compression of the lungs, making

it difficult for cats to breathe. It is a symptom of a number of different diseases, including heart disease, infection, FIP, tumours and diaphragmatic hernias. Symptoms include anorexia, respiratory distress, abdominal and mouth breathing and cyanotic mucous membranes. Diagnosis is by ultrasound or X-ray, but sometimes the vet will diagnose by performing thoracocentesis just based on the cat's presenting signs. Treatment involves draining the fluid, sometimes via the placement of a chest drain. The prognosis depends on the cause but requires rapid initial treatment. Pleural effusions are normally caused by serious disease.

42 *Pododermatitis* – this is caused by bacteria, tumours or a number of other dermatological diseases. The condition results in swelling of the pads of the feet, lameness and discomfort, and systemic signs including pyrexia, lethargy and anorexia. Diagnosis requires a rigorous and extensive workup, as there are so many potential causes, but biopsies may be required. Treatment includes antibiotics and anti-inflammatories but largely depends on the underlying cause, as does the prognosis. Some cases resolve spontaneously.

43 *Renal failure (acute)* – acute renal failure (ARF) can occur as a result of an infection, toxin or blockage within the urinary tract. Diagnosis is made by blood tests (which will show azotaemia) and urinalysis, but unlike chronic renal failure there is unlikely to be any significant weight loss, polydipsia or polyuria. Treatment requires aggressive fluid therapy and treatment of the underlying cause, if known. Some cats with ARF will stop urinating, in which case

diuretics are administered. Prognosis depends on the underlying cause but, if this can be treated and urine output maintained, prognosis is good.

44 *Renal failure (chronic)* – chronic renal failure (CRF) is a condition usually seen in older cats. Symptoms include polyuria, polydipsia, decreased appetite, weight loss and dehydration. Diagnosis is made via blood tests that show azotaemia and urinalysis showing dilute urine. Treatment usually involves a change in diet to a reduced-protein, low-phosphorus one, and other treatments according to the cat's specific needs – e.g., potassium supplements, appetite stimulants and anti-sickness drugs. If the cat becomes inappetant and very dehydrated, a drip may be put in place for a period of time to attempt to diurese the kidneys; however, this is only a short-term treatment. Some vets will subsequently recommend subcutaneous fluids that can be administered by the owner at home. Prognosis depends on how quickly kidney function is deteriorating and this varies enormously. Ultimately a decision will usually have to be made to euthanase the cat when quality of life is no longer acceptable.

45 *Ringworm* – this is a skin infection caused by a fungus. It is contagious to cats and humans. Symptoms vary but can include circular patches of hair loss, scaliness, skin infection and pruritis. Diagnosis can include examination of the cat with an ultraviolet light (as some types of the fungus fluoresce) and microscopic examination of hairs and fungal culture. Treatment involves tablets and baths with medicated shampoo. In the case of longhaired cats it can be helpful to clip them all over. Any other animals in

the same environment should be checked for infection and the house and associated areas will also need decontamination with appropriate disinfectants. The prognosis is good but long treatment is often required and the resolution is harder the more cats there are within the household.

46 *Seizures* – these are due to abnormal electrical activity within the brain. They can display as a typical fit with spasmodic leg movements, inability to walk, defecation, and salivation; a partial fit affecting one limb, face or head; or even as a strange walk (moon-walking) or vocalization. Causes include epilepsy, head trauma, tumours, hypertension, infection, toxins and hypoglycaemia. The underlying cause is diagnosed via blood tests, analysis of the fluid in the spinal cord (cerebrospinal fluid) and MRI scan. Treatment of epilepsy, which is usually initiated only if seizures become frequent or severe in nature, will involve tablets. If seizures develop that do not stop within a few minutes, or your cat has repeated seizures within a short space of time, it is important a vet is contacted as a matter of urgency.

47 *Squamous cell carcinoma* – this is a type of cancer that is seen most commonly in cats with white ears and noses, especially sun worshippers. Crusting of the ear tips, eyelids and nose is initially seen. This can quickly spread and may metastasize. This type of tumour is also seen orally under the tongue but not just in white individuals. When the ears are affected, treatment involves amputation of the ear tips. Eyelid or tongue tumours cannot be removed in this way, although nasal surgery can be helpful for some nose

tumours. Prevention involves application of sunblock on thinly furred white areas. The prognosis is good if full resection can be achieved on the ear tips. If on the nose it will invariably spread and ulcerate and ultimately euthanasia will be required, but this may be some time after the initial diagnosis. Radiotherapy may help but is not readily available and some vets use cryosurgery (using extremely low temperatures to freeze and destroy tissue). If the tumour is in the mouth, euthanasia will be necessary when the cat has difficulty in eating or quality of life is poor.

48 *Tail pull injury* – this is commonly seen in a cat that has been involved in a traffic accident or a cat that has had its tail trapped in a door and attempted to pull it out. It results in separation of the vertebrae and damage to the nerves that emanate from that portion of the spinal cord. Cats suffering from this may have concurrent injuries but will also have a paralysed tail caudal to the pull. Diagnosis is based on symptoms and X-ray. The major problem with this injury is that the pull may cause damage to the nerves that run to the bladder meaning that the cat is unable to urinate. Sometimes this function returns with time, in which case treatment involves amputation of the tail to prevent further damage. If the cat is unable to urinate short term the bladder can be expressed manually; however, longer term, this dramatically affects the quality of life of the cat and euthanasia has to be considered. The prognosis depends on the severity of the pull.

49 *Thromboembolism* – this is due to production of a blood clot within the circulation and is often associated with heart

conditions. Cats present in acute-onset severe pain. The symptoms depend on where the clot lodges, but the most common site is where the aorta branches off to supply the hind legs. If this is the case, the cat will be unable to walk on its hind legs, the feet will become cold and there will be no pulse in the hind legs. Sometimes the clot lodges just in one leg. Other sites include the brain, lung and kidney. Diagnosis is usually based on the symptoms, as it causes such extreme discomfort for the cat. Treatment can be attempted by giving blood-thinning drugs and analgesia, but it is rarely successful and can be very distressing for the cat. If treatment is successful the problem frequently recurs, as there is usually an underlying condition. The prognosis is very poor.

50 *Urolithiasis* – this is due to the production of crystals in the cat's urine. These crystals can merge to form stones in the bladder. Symptoms include pollakiuria, dysuria, inability to urinate, haematuria and excessive grooming of the perineal and abdominal areas. Diagnosis is by ultrasound and X-rays, including contrast radiography. Treatment depends on the type of stone present and its size. If the crystals are small and of a particular type, then specific diets will alter the pH of the urine and they will dissolve. If the crystals are large or non-dissolving, then surgery may be required to remove the stones from the bladder. The prognosis is good, although this problem can recur without appropriate dietary and lifestyle management.

APPENDIX 3

Useful Websites

THE INTERNET IS AN AMAZING TOOL BUT NOT ALL information found on it is correct. You can easily get extremely confused if you read advice on a website and your vet tells you something completely different. Websites are not regulated, so anyone could post information that is basically nonsense and dress it up to look professional. Listed below are those websites that I personally feel are worth going to regarding cat matters. There are probably no enquiries that these websites, one way or another, cannot answer!

If you don't have access to the Internet you could visit your local library, where you can request assistance in viewing these websites. All of them provide contact information with addresses for correspondence and telephone numbers for those of you who prefer a more traditional means of communication.

Animal Health Trust: www.aht.org.uk
This is an excellent website for frequently asked questions and fact sheets.

Association of Pet Behaviour Counsellors: www.apbc.org.uk
Visit the APBC website for details of your nearest pet-behaviour counsellor and information about choosing this profession as a career.

Blue Cross: www.bluecross.org.uk
They have some great information sheets.

Cat Chat: www.catchat.org
This charity's site is all about rescue information and much more.

Sarah Caney: www.catprofessional.com
Sarah is a great friend and an absolutely wonderful and highly respected cat vet. You can download from her website excellent e-books on specific diseases. She manages to make complicated medical stuff easy to understand.

Cats Protection: www.cats.org.uk
They have great fact sheets; this website is a good source of information.

Celia Haddon: www.celiahaddon.co.uk
Celia is another great friend and excellent source of cat advice. She has numerous articles about cats on her website, where you can get important advice.

DEFRA: www.defra.gov.uk
You should visit this site for any information about travelling with your cat and for advice regarding emigration.

Battersea Dogs and Cats Home: www.dogshome.org
This rescue charity's site has some advice sheets together with details of their behaviour help line and a dedicated help line for lost dogs and cats.

Feline Advisory Bureau: www.fabcats.org
I don't think you can beat this charity for clarity of information and enthusiasm. If you don't find the answer on their website, for a small donation you can get an expert in the field to answer your specific enquiry.

GCCF: www.gccfcats.org
This is the official website for the Governing Council of the Cat Fancy. It has lots of information about pedigrees and showing cats, if that's your interest.

PDSA: www.pdsa.org.uk
The People's Dispensary for Sick Animals has more good fact sheets and a useful site should you ever have any concerns about affording vet bills.

Pet Health Council: www.pethealthcouncil.co.uk
This is another site with good leaflets on general cat care.

Pet Food Manufacturers Association: www.pfma.org.uk
There is good nutritional advice on this site.

RSPCA: www.rspca.org.uk
There is a great deal of information on this site about all species, the work of the RSPCA in general and their current campaigns.

Scottish SPCA: www.scottishspca.org
This Society for the Prevention of Cruelty to Animals site is the one for you if you live in Scotland.

Vicky Halls: www.vickyhalls.net
Obviously I can't complete this list without including my own website!

INDEX